DRONE
WARFARE

DRONE WARFARE

The Development of Unmanned Aerial Conflict

Dave Sloggett

Skyhorse Publishing

Skyhorse Publishing books may be purchased in bulk at special discounts for sales promotion, corporate gifts, fund-raising, or educational purposes. Special editions can also be created to specifications. For details, contact the Special Sales Department, Skyhorse Publishing, 307 West 36th Street, 11th Floor, New York, NY 10018 or info@skyhorsepublishing.com.

Skyhorse® and Skyhorse Publishing® are registered trademarks of Skyhorse Publishing, Inc.®, a Delaware corporation.

Visit our website at www.skyhorsepublishing.com.

10 9 8 7 6 5 4 3 2 1

Library of Congress Cataloging-in-Publication Data is available on file.

Cover design by Jon Wilkinson

Print ISBN: 978-1-63220-505-6
Ebook ISBN: 978-1-63220-874-3

Printed in the United States of America

Contents

Preface

Almost since the dawn of aviation, engineers and scientists have mused upon and sometimes developed aircraft that could be controlled remotely and therefore fly unmanned. This book charts their evolution and the enabling technologies that made their development possible. It does this over a period of nearly 100 years, from its early beginnings through to the period after the Second World War when applications of such platforms quickly grew to the plethora of modern-day military and civilian uses to which they are now routinely applied.

That history is punctuated by a series of technological developments that have allowed unmanned aircraft to gradually emerge from the shadow of the manned platform. The first driver for the development of unmanned aircraft came from the military. They could see the potential, even in the early stages of the evolution of manned air platforms. Since then, a number of operational drivers have also emerged that have had a catalyzing effect upon the development of unmanned aircraft. Not least among these is the development of asymmetric warfare and counter-insurgency operations.

The initial challenges lay in finding ways to control the platform, both on the ground and in the air. With the development of the gyroscope and greater understanding of the dynamics of flight, solutions were found to controlling unmanned aircraft. However, these came too late to have a meaningful impact on the First World War.

In the interwar years the level of investment in unmanned aircraft did not match that going into manned platforms. That is apart from in the days leading up to the Second World War in Germany where tentative steps were being taken that would see the emergence of the V-1 flying bomb as the first genuine military application of unmanned aircraft. Its next genesis would have to wait until it was possible to remotely control an unmanned aircraft, albeit with restrictions regarding line of sight.

The mainstream media likes to characterize the contemporary guises of these platforms by the generic term 'drones'. The origin of this use of language is multi-faceted. It is possible to suggest that one derivation of the term dates back to the application of unmanned aircraft as targets for various gunnery and air-to-air missile tests. Drones at that time were seen to be simple unarmed devices whose flight times were short in duration. One example was the Jindivik that was developed by the Australian Government Aircraft Factory (GAF) as a result of a joint programme between Australia and the United Kingdom. Its name derives from the Aboriginal word meaning 'the hunted one'.

The next stage in the evolution of unmanned aircraft occurred when advances in technology allowed them to start to carry sensor systems in addition to the basic payload required to control the platform. Pioneering work in Israel was quickly picked up by the United States. This initial work has seen a plethora of platforms built to provide eyes and ears on an adversary's actions, both in the land and maritime environments. As technological developments occurred, so the larger of the platforms could also begin to carry weapon systems. The kind of asymmetric conflicts occurring in Iraq and Afghanistan helped create drivers for yet more technological advances in the capabilities of unmanned aircraft. Mission endurance in particular was a huge issue.

The longer an unmanned aircraft could stay on station, the more it could help to build up intelligence on the pattern of life in an area. As the need for the military to apply precise effects on the ground increased, so grew the need to maintain a platform on station. With many people's behaviour linked to diurnal cycles it was natural that platforms should be developed that are capable of remaining on station over a 24-hour period. Developments in satellite communications technology have also provided a crucial enabler allowing missions conducted by unmanned aircraft to be extended.

At the time of writing, unmanned aircraft are still controlled remotely by human beings but they are reliant on satellite communications to maintain the control. If the satellite links become degraded for a period of time, the unmanned aircraft has to have a limited decision-making capability. If, for example, satellite communications are lost, the unmanned aircraft will initiate a return to its home base. With an increasing geographic spread of operations comes an increased need to operate under a range of climatic conditions. Just because a Reaper on a mission over Pakistan decides that it has lost satellite communications and should therefore return to its main operating base does not necessarily mean the conditions there are suitable for a landing. This provides evidence that we are some way from being able to think of drones as fully autonomous devices.

Some contemporary media writers like to use the term 'drone' because it carries with it a sense of the rampant application of technology. They seek to portray these platforms as a form of Frankensteinian monster created by mad people who will one day rue its invention. This image is one that anticipates the next stage in the development of unmanned aircraft, when the link with human control is either partially or completely severed. Citing the degree of autonomy already available to platforms such as the Predator and Reaper, in the case of them suffering some form of mission failure the naysayers argue that this is the thin edge of the wedge and that fully autonomous robotic unmanned aircraft are just around the corner.

The use of the term 'drone' is subtle and implicitly derogatory. It evokes images of the classical genre of science-fiction movies that have tried to show the inevitable conclusions when man hands over control to robots. Of all these caricatures, the sentient computer – HAL 9000 – in Stanley Kubrick's famous film *2001: A Space Odyssey* is perhaps the most alarming. In the film HAL has been programmed to achieve a mission. When the crew of the spaceship appears to threaten that objective, the computer takes what it believes to be the logical step of removing them from the picture.

For those seeking to add a negative hue to the portrayal of unmanned aircraft this provides the perfect illustration of what is essentially a moralizing message. If man leaves decision-making to robots (whatever that actually means), the outcome will be bad. However, not all Hollywood depictions of robots have been alarmist. For example, in *Star Wars* two of the main characters have comedic qualities.

In the end, according to this narrative, what you get is a fully-functioning form of exterminator that operates entirely autonomously, irrespective of the implications for mankind. A world fully or partially dominated by robots is a theme to which Hollywood has returned on a number of occasions and the language accompanying this portrayal of drones is almost entirely negative.

Others in the media use it to portray a quite different viewpoint, citing the role of the drone bees in a hive going about their business without a great deal of thought. The implication is that the drones are servile and verging on automata. In fact, this is a weak metaphor. Any experienced beekeeper will tell you that the complexities of the operations of the hive are far more dynamic than some might believe. If the queen bee is a feisty character, the drones will pick up her mood. This can create quite distinct behaviours in hives located within a few metres of each other. To date, no unmanned aircraft has developed the capability of expressing the kind of moody emotions adopted in a hive ruled by a difficult queen bee.

The problem with these portrayals of unmanned aircraft as some form of autonomous killing machines is that it does not accurately reflect the reality on the ground. While some scientific research is being undertaken into the ways in which unmanned aircraft can be given increased levels of autonomy, the idea that they will soon be self-selecting targets is somewhat fanciful. Developments in the artificial intelligence area have not yet reached a level of maturation where such decision-making on a rational basis is possible by a machine viewing what is often a dynamic and complex environment. Does anyone touting this narrative actually believe that many of today's political leaders would really hand over such decision-making to software? It is very doubtful, given all the careful measures he has taken to rein in and structure armed attacks over Pakistan, that President Obama would sign off on letting the unmanned platform make its own killing decisions.

However, as the capabilities of the various unmanned platforms grew, military people in particular became irritated by the ways in which the media was manipulating the language surrounding the capability. Unlike those members of the press that liked to carp on about drone technologies, military people saw their utility. In many cases they have saved lives on the ground, forewarning of an ambush or the presence of an IED (improvised explosive device).

To the military users of drones, the media caricature was simply wrong. It was important to develop a more rational set of descriptors for the platforms. As ever in the military world, this led to the proliferation of a new set of three-letter acronyms. These include Remotely Piloted Vehicle (RPV), Unmanned Aerial Vehicle (UAV), Unmanned Aircraft (UMA) and Remotely Piloted Aircraft (RPA). Each tried to portray a subtle difference between the capabilities of the various platforms that were emerging. For this book the term Unmanned Aircraft (UMA) has been adopted as a generic descriptor covering all the variants that have either been developed or are part of ongoing research activities.

At the start of the twenty-first century the topic of UMA is the subject of a great deal of media attention, most of it critical. The Predator UMA is an aircraft that has received a great deal of public and private scrutiny. Questions about their operations abound in the media. Are they lawful? Do their operators somehow detach themselves from reality and imagine that they are involved in some sort of computer game? How do they deal with what can be a monotonous surveillance task? Does this make them more ready to engage a target quickly? What are the psychological effects on the operators who go home to their families after a day at war? How do they decompress? Who authorizes the missions and what steps are taken to avoid civilian casualties? Where does all of this end up? Do we really need to retain this capability?

UMA are not confined solely to the air domain. At sea, both above and below the water, unmanned platforms will have a role in the application of military power, in whatever form it may take. Tentative steps have also been taken in space with the launch of a smaller unmanned version of the space shuttle.

In the maritime domain the problems posed by pirates off the coast of Somalia required some form of persistent observation capability over large areas of the Indian Ocean. With increasing worries about the growth of piracy in the Gulf of Guinea, it may not be long before unmanned platforms are also flying in support of local naval forces trying to intercept the pirates before they conduct their attacks. In Indonesia where the problems of pirate attacks at the geographically-dispersed anchorages challenge local coastguard and naval assets, UMA could well provide the kind of persistence required to help eradicate attacks on merchant vessels awaiting entry into ports.

Using UMA against pirates is only one of a wide range of maritime security activities. Interdicting drug-smuggling routes in the Gulf of Mexico and economic migrants trying to reach Australia from Indonesia highlight other applications of unmanned platforms. With some unmanned aircraft now flying in the upper reaches of the atmosphere, it is even possible to suggest that they may at some point become involved in confrontation on the edges of space.

This may all sound a little fantastic, even far-fetched. However, the simple fact is that what was once seen as existing only in the world of science fiction has rapidly made the transition into science fact. Furthermore, despite the negative coverage that armed UMA receive in some quarters of the mass media, they are simply not about to go away any time soon.

To address this topic comprehensively, the book is structured into eight chapters. One thing it does not do is provide a catalogue of all the various armed and unarmed UMA that have been developed. There are other books referenced in the bibliography that act as a fine reference for those needing to know who has developed which drone.

After an introduction that describes the background to the role played by UMA in current military operations, the second chapter lays out the foundations on which initial interest in UMA emerged. The chapter explores the early days of unmanned flight and the innovations in remote control that helped lay the foundation for future development. A theme of using UMA to deliver a warhead against a military target was to be the primary application of the technology for its first thirty years. The development of the aerial torpedo was the embodiment of this concept.

During the Second World War that capability found its genesis in the development of the V-1 flying bomb. Chapter Three charts the first serious uses of UMA in warfare, noting the development in Germany of the V-1. A detailed comparison is also provided of the impact of the V-1 with what happened during the Blitz.

Chapter Four examines the initial development of new uses of UMA in the immediate aftermath of the Second World War. Chapter Five notes the burgeoning applications of UMA that occurred around the time of the Vietnam War which breathed new life into the development of UMA and laid the platform for its more recent incarnations. Those chapters complete the historical context against which the current uses of UMA have developed.

From that point onwards the content of the book focuses on the current and future uses of UMA. Chapter Six looks at the contemporary application of UMA and specifically their impact in Pakistan. This chapter presents a unique and previously unpublished view on the psychological impact of the armed UMA strikes that challenges perspectives suggesting they have a radicalizing effect upon the affected population. The analysis presented here is backed up by detailed studies in three later Appendices.

Chapter Seven looks to the future of unmanned aerial conflict, noting some of the technological barriers that currently inhibit further development of UMA in the military sphere. The discussion also briefly ventures into the civilian applications of UMA, such as in the area of disaster relief and the fight against transnational organized crime. The final chapter then looks back and draws some conclusions from the preceding analysis. The aim is to provide the reader with a detailed historical timeline of the developments of UMA but also to challenge in a balanced and analytical way certain myths that have arisen along the route.

Acknowledgements

I am a child of the Cold War. At 10 years of age the Cuban Missile Crisis had a profound impact upon my view of the world. Mankind had teetered on the edge of an abyss and somehow contrived to pull away from Armageddon. At the time, of course, I had simply no idea of what was at stake.

I first went to work at 16 years of age at the Royal Aircraft Establishment at Farnborough. Within weeks I was involved in developing research activities that would see me visit many Royal Air Force stations in the United Kingdom. In the course of my work I was also fortunate to visit Iceland and Norway on several occasions. What was a geographical focus on the north in the early part of my career was to change significantly as I entered its twilight stages. More recently my geographic focus has turned eastwards.

In those early days one of the main focal points for the defence of the United Kingdom was in the north, where long-range Russian bombers or naval vessels could launch pre-emptive attacks against airfields. I was to take many lessons away from my early years at Farnborough into a career that has now spanned over forty years.

More recently it has seen me making visits to Kosovo, Iraq, Afghanistan, the Indian Ocean and West Africa. The threat to the United Kingdom is now more diffuse; its magnetic polarity less certain than during the Cold War. In those overseas theatres of war I have been fortunate in working with people who have helped me understand the very different operating environment in which conflicts are now conducted.

I am therefore really grateful to Pen & Sword for the opportunity to produce this book. My thanks go to Martin and John for their support in getting this project set up and to Matt for all his help and encouragement. I also need to thank my friends in *IHS-Jane's* for the opportunities to publish related articles and to explore some of the arguments while researching its content.

I am also specifically indebted to *IHS-Jane's* for access to their Terrorism and Insurgency Centre database of nearly 200,000 records that document the activities of insurgents and terrorists across the world. Will Hartley and James Green have both been very supportive in my efforts to access and analyze the data. Other datasets that have been a source of important insights include

ACKNOWLEDGEMENTS

those published by the New America Foundation, the Long War Journal and the International Bureau of Investigative Journalism. Slight inconsistencies in their records of how many people have died in armed UMA attacks in Pakistan, Yemen and Somalia are simply down to the problems of accessing accurate data on the ground and a lack of transparency from the White House. I would also like to acknowledge the data collected by the South Asia Terrorism Portal on attacks on NATO tankers in Pakistan. Information on the operations of UMA in Afghanistan has been harder to come by from open sources and therefore is available in slightly less detail.

The bibliography in the back of this book provides a guide to the main sources I used in compiling the text. Special mention must go to the detailed work of Yefim Gordon who is a prolific writer on Soviet, Russian and Chinese aviation. His writings on all aspects of aviation have been an inspiration.

As I embarked upon this journey I have benefited hugely from the guidance and support of a number of military officers with whom I can honestly say it was a pleasure to work. Brigadier Chris Holtom stands out as a mentor in my early years. Every conversation I ever have with Chris leaves me awestruck about the depth of his understanding of the world in which we all now live. He was the first person in the British military I ever met who really understood the impact of societal landscapes. He is a kindred spirit.

In pursuing this subject I have been able to draw upon my experiences of visiting a number of operational theatres. I am grateful to those, some of whom have to remain unnamed, for their support and for making each visit such a pleasure. My time in Kosovo was particularly instructive and shaped my awareness of the complexities of what I call societal landscapes: the people, their societies and their rules, their customs, beliefs and traditions that govern behaviour.

It was during the course of several visits to Kosovo and my subsequent visits to Iraq that I learned of the multiple agendas that can fracture and divide societies along ethnic, religious and criminal fault lines. In looking at these societies and their underlying complexities I was fortunate to have been able to draw upon the understanding I developed of the nature of societal division that existed in Northern Ireland, albeit at a somewhat simpler sectarian level of motivation. These overseas visits laid the foundations on which I draw in writing this work.

I was fortunate to be able to carry that understanding forward into working with Iraqi military people through NATO as they sought to take control of the security of their country. Several of those with whom I worked have since perished in the ongoing struggle in Iraq. I lament their passing. My lectures often provoked intense debate but this always took place with a high degree of civility and respect for each other's viewpoints and perspectives.

There are also times when you meet quite exceptional individuals who, day in and day out, put themselves in harm's way. To Nick, Paddy and the rest of the guys with whom I have worked both in theatre and in the United Kingdom I would like to offer my sincere thanks for being such an inspiration.

A specific 'thank you' goes to Andy for carrying my bags and looking after me in Iraq. It was there that I saw a deeper impact of the fault lines that can divide societies. In Afghanistan my understanding of these issues developed to another degree. Recent visits into West Africa have also provided another calibration of that understanding of the ways in which societal landscapes differ.

Brigadier Paddy Allison also deserves a really big 'thank you' for being such an active supporter of my research efforts in the early days. I must also take this opportunity to thank my friends in MI Brigade and the Allied Rapid Reaction Corps, both of which I have worked with on my recent journey. It was a specific pleasure to work alongside Brigadier Iain Harrison in the build-up to the ARRC's deployment into Afghanistan. It was also good to stay in touch with him while he was in theatre.

I would also like to acknowledge the privilege I had of meeting and working with Major General Andrew Mackay and his team from 52 Brigade in the period prior to their deployment to Afghanistan and during my visit to theatre. To have been able to sit in on the planning meetings before the operation to free Musa Qal'eh was a fantastic experience.

My visit into Afghanistan in 2011 to witness the operations of the American and United Kingdom's fleet of unmanned aircraft at first-hand was also really informative and helped direct my subsequent writing efforts. For that visit I have to thank my friend Air Chief Marshal Sir Stephen Dalton. The Royal Air Force has been hugely lucky to have been led by a man with such knowledge, mental agility and who is so approachable. I always leave our discussions much the wiser. He has been a fantastic guide and mentor. I was honoured to be asked to give a short address at the annual Battle of Britain Dinner that was hosted by Sir Stephen Dalton. I will never forget that evening. The image of a lone Hurricane flying low over the airfield at RAF Northolt is one that is seared on my memory. It brought a lump to my throat and tears to my eyes.

A specific 'thank you' also needs to go to Brigadier General Herbert

ACKNOWLEDGEMENTS

McMaster in the United States who has always encouraged my early efforts at writing about the problems of counter-insurgency operations and to my good friend Simon who somehow manages to fit in reading my papers and providing positive comments, drawing on his own extensive experience on active service.

I would also like to specifically acknowledge the help of my second son Chris in researching the subject on the legal matters associated with the operations of unmanned aircraft and their use in asymmetric warfare. This was a complex area to fathom as it involves many legal issues that in a balanced treatment need to be explored in some detail. Chris did what any Oxford history graduate would do and applied himself diligently to the subject.

My wife Jo also continues to be my inspiration, interrogator, editor and supporter. Her encouragement and belief is the mainstay of my efforts. This book, as ever, is dedicated to her and my three sons with a huge amount of love and affection.

Dr Dave Sloggett
Ryde, Isle of Wight
July 2013

Terminology

Anyone involved in the field of unmanned aircraft knows that there is a great deal of debate surrounding the nomenclature associated with the various forms such platforms can take. A plethora of terms has emerged to try to cover all the different possible configurations that an unmanned aircraft can take. One of these is the Unmanned Aerial Vehicle; a term used in NATO circles.

Being precise about the definitions is increasingly important as some are beginning to appear as specific terms in legislation that defines what unmanned aircraft can do when they are operated in managed civilian airspace. Already a number of air misses over Iraq and Afghanistan have shown the potential problems that might arise in the future if the current rapid rate of development of civil applications of unmanned aircraft continues.

The media, of course, likes to refer to unmanned aircraft as drones. In the book I explore this use of language and its motivations. While the term is a military shorthand used at the tactical levels of command, this is not something welcomed by senior military commanders. The name 'drone' carries with it the connotation of somehow not being under control; something that senior military commanders simply do not wish to convey.

In this book I have opted for the use of the broad term Unmanned Aircraft (UMA). In the United Kingdom, terminology developed by the Defence Concepts and Doctrine Centre at Shrivenham makes it clear that a UMA is 'an aircraft that does not carry a human operator, is operated remotely using varying levels of automated functions, is normally recoverable, and can carry a lethal or non-lethal payload'. Beyond the aircraft itself, the ground component is recognized in a wider definition as an Unmanned Aircraft System (UAS): this embraces both the airborne and ground-based elements of control and mission exploitation when intelligence data is collected.

This definition reveals some of the issues that have emerged in the growing lexicon associated with UMA since their inception. In the book I occasionally use the widest possible term 'platform' as a generic catch-all for all forms UMA can take, irrespective of whether they are radio-controlled or rely on internal navigation systems. Problems have arisen when UMA have been adapted to be able to be manually flown for part of the mission: a technique used in the Soviet Union in the early days when the pilot then ejected before the UMA was destroyed by an intercepting aircraft.

Abbreviations

AEW	Airborne Early Warning
AMRAAM	Advanced Medium-Range Air-to-Air Missile
AQAP	Al Qaeda in the Arabian Peninsula
BAMS	Broad Area Maritime Surveillance
BVR	Beyond Visual Range
CIA	Central Intelligence Agency
COBRA	Coastal Battlefield Reconnaissance and Analysis
COIN	Counter Insurgency
COMINT	Communications Intelligence
DARPA	Defence Advanced Research Projects Agency
ELINT	Electronic Intelligence
EUROPOL	European Police Organization
EW	Electronic Warfare
FATA	Federally Administered Tribal Areas
FEBA	Forward Edge of the Battlefield Area
HALE	High-Altitude Long Endurance
HUMINT	Human Intelligence
IAF	Israeli Air Force
IDF	Israeli Defence Force
IED	Improvised Explosive Device
IMINT	Image Intelligence
ISTAR	Intelligence, Surveillance, Target Acquisition and Recognition
JATO	Jet-Assisted Take-Off
JDAM	Joint Direct Attack Munition
LCS	Littoral Combat Ship
LRS-B	Long-Range Strike-Bomber
MEZ	Missile Engagement Zone
MOD	Ministry of Defence
MUAS	Maritime Unmanned Air System
NATO	North Atlantic Treaty Organization
NRO	National Reconnaissance Office
NWFP	North West Frontier Province
ORBAT	Order of Battle

RATO	Rocket-Assisted Take-Off
RWR	Radar Warning Receiver
SEAD	Suppression of Enemy Air Defence
SIGINT	Signals Intelligence
TALD	Tactical Air-Launched Decoy
UAS	Unmanned Air System
UAV	Unmanned Air Vehicle
UCAS	Unmanned Combat Air System
UCLASS	Unmanned Carrier-Launched Airborne Surveillance and Strike System
UMA	Unmanned Aircraft
UOR	Urgent Operational Requirement
WAPS	Wide Area Persistent Stare

CHAPTER 1

Prologue

This is a largely fictitious account of the last journey of Ilyas Kashmiri to Leman in South Waziristan that is written to provide an insight into the ways in which armed unmanned aircraft are used to track terrorists and insurgents. While this is very unlikely to be accurate, it has been written after a careful analysis was undertaken of the material that entered the public domain in the wake of his death on 3 June 2011.

Ilyas Kashmiri must have been a slightly worried man as he boarded his Toyota Land Cruiser and headed south. Ahead of him was a difficult journey to his new base in South Waziristan. He would only reach that by running the gauntlet of drones that seemed to be forever patrolling the skies over his current hideaway. For him death would emerge suddenly from the sky with little warning; a bolt literally from the blue. But that was a risk he welcomed. As his colleagues in the vehicle knew, he despised the Americans. Let them do their worst. Every time the drones killed one of those who were fortunate enough to do Allah's work, more quickly stepped in to take their place.

Ilyas Kashmiri had taken refuge in Mir Ali in the immediate aftermath of the attack by Pakistani security forces that had ended the siege of the Red Mosque in Islamabad in July 2007, joining up with Taliban and Al Qaeda fighters operating in the area. It was a place he quickly grew to love; dawn and sunsets there could be especially beautiful, if not stunning on occasions. This was a location in which several insurgent networks were operating. Abu Kasha al Iraqi, a close associate of the leadership of Al Qaeda, also lived in the area. But with so many potential targets living in close proximity to each other, a spate of armed UMA strikes towards the end of 2010 had made Ilyas Kashmiri re-think his options. He had concluded that Mir Ali was no longer safe. The accuracy of the strikes had also puzzled him. How could the Americans know the location in which many of his colleagues had gathered? Was the CIA operating a network of informants in the area?

Kashmiri's journey through life had provided him with a wide range of skills. At 45 he had curriculum vitae that would have been the envy of many of his counterparts involved in the insurgency. He lost an eye fighting Russian forces in Afghanistan in the 1980s and had escaped from an Indian jail where he was imprisoned for two years, having been captured in Kashmir. His group, the 313 Brigade as it was known, had been banned in the backlash against terrorism after the 11 September attacks in America. Demonstrating a certain stubborn streak, Kashmiri had refused advice from Pakistan's Intelligence Service, the famed ISI, to join his 313 Brigade with Jaish-e-Mohammad, a group whose focus was on Kashmir. He was determined not to be shackled by any connections that he did not select himself.

After being arrested by the Pakistani security forces for having been allegedly involved in an attempt on President Musharraf's life in December 2003, he was released because of a lack of evidence. His connections deep inside the ISI had really helped him on that occasion. This was not the only occasion when he was to spend some time in one of Pakistan's jails. In 2006 he had come to the attention of the Americans when they suspected his involvement in an attack on the American Consulate in Karachi. Each of these steps helped to build his notoriety and establish him as a key player in the insurgent leadership. He saw these additions to his combat curriculum vitae as blessings.

He had also recently been cited by the traitor David Hedley who had arranged a plea bargain with the United States Justice Department to avoid the electric chair when Hedley accused Kashmiri of asking him to look into the potential for attacking a leading United States defence contractor. The motivation for this attack was alleged to be Kashmiri's anger at the drone strikes.

In 2010 he caused alarm in several European capitals when remarks attributed to him suggested that he had personally already sent a number of terrorist teams into Europe. He was pleased with the result. It showed how seriously people took him. It enhanced his position in the insurgency.

Using people who had been trained in Pakistan and held valid European passports, Kashmiri knew it would be possible to send certain individuals home equipped with the knowledge of how to build bombs and carry out attacks. The presence of a number of people of German and British origin in the area lent credence to his claim. Two brothers from the Midlands in the United Kingdom were killed by a drone strike as a direct result of his remarks. Kashmiri's status as a key player in the Al Qaeda network of groups was now quite clearly established. He was on the radar horizon of western intelligence agencies.

PROLOGUE

For Ilyas Kashmiri, and many others like him, North Waziristan had recently become a distinctly hostile environment. Nearly every day the sound of at least five drones – or maybe more, it was so hard to tell – permeated his hearing. He found it hard to switch off, being unsure if at any point one of them could have been the harbinger of doom. Local people likened the noise made by the engine of the Predator to that of a wasp. For many people in the area of Miranshah the sting from the drones had proved fatal as the rate of attacks mounted by the Americans seemed to increase almost daily. Kashmiri had heard rumours of at least forty strikes in the last four months.

Setting off from his base close to Mir Ali, 25 miles east of Miranshah, the main town in North Waziristan, the journey to his new base in Laman was going to take a few days. He settled back into the rear seat of the Land Cruiser and looked out across the green plains of the Tochi River. Summer was upon them now and the river was at its lowest for some time but the complex series of drainage ditches dug by the locals still provided enough water for their crops to grow. As they drew into a small village the air became heavy with the scent of *Thymus serpyllum*, also known as 'creeping thyme'. Its lilac flowers were in abundance in the area. Locals in the region burned it for medicinal purposes. The herbal nature of the smell briefly drove Kashmiri into a reverie. He longed for the scent of musk that would signal his entry into Paradise and eternal joy. His thoughts were then interrupted as the car lurched over a divot in the road and he quickly focused again on the matter in hand.

For Kashmiri, Mir Ali and the local environs seemed to have become a new locus of drone attacks in the past few weeks. The attack on 16 May was far too close. He had lost four friends that day to the sting of the wasp. The single Hellfire missile had targeted a compound literally a few hundred metres away from this own hideaway from where he had plotted the attacks in Mumbai and the recently highly successful 17-hour-long occupation of the Pakistani naval base in Mehran near Karachi. The results of that attack had particularly pleased him as it showed just how inept the Pakistani military were when confronted on their own military bases. He had similarly enjoyed the discomfort of the Indian security authorities during the sixty-two hours of the Mumbai siege in November 2008.

He had intended the attack at Mehran to be part of a series of moves to avenge the death of Osama bin Laden. This was simply the first. More were to come. But to carry those out and truly avenge the death of his leader and friend, he needed to survive. The move south, into a new operating area far away from the sound of the drones, was imperative.

These drones were more than annoying. Since the start of the New Year he estimated that over forty strikes had taken place in and around Miranshah. The Americans seemed to be zeroing in on his location and that had spooked some of his colleagues who had urged him to make a move south.

Kashmiri was initially reluctant to make the move. He did not fear death. In fact, his desire for a glorious martyrdom drove his very being. So many had been able to achieve it before him; when would he join them in Paradise? His colleagues had prevailed upon him to move. The threat of an onslaught on the area by Pakistani ground troops also preyed on their minds. In the end it was not too difficult a choice to make. It was only a question of where their new headquarters was to be located.

South Waziristan offered an attractive solution. It was not too far to travel, reducing the time Kashmiri would be exposed to the ubiquitous drones. With any luck if they set off at the right time, blending into the busy early-morning traffic streaming along the river valley, their journey would not stand out. The area was always busy in the early morning as people either travelled to market or to the main manganese ore excavation sites at Razmak to the north-west, a place favoured by many in the local population as a holiday resort.

His departure from Mir Ali came with mixed feelings. It had been particularly pleasant living in the hilly tracts or Khaisora of the region, sitting out under the shade of the apple trees sipping tea and talking of jihad and its everlasting rewards in Heaven. His Waziri hosts had been most hospitable, fully in keeping with their Pashtun traditions and the creed of Pashtunwali. But even their leaders were beginning to feel the strain of the drone attacks. They simply did not know when or where the next one would originate. It was draining.

Moving to Laman near the important town of Wana in South Waziristan provided Kashmiri with an opportunity to start afresh, away from the prying eyes of the drones. As he thought about the new opportunities that would arise in Wana, he smiled. There was a certain delicious irony in that Wana, a centre for British forces combating an insurrection led by the Karlanri Tribal Confederation of Waziristan in the 1930s should become his base some eighty years later from which to develop and plan attacks against the United Kingdom. He very much hoped that history would repeat itself but with a slightly different outcome. He was very much looking forward to joining up with the members of the Ahmadzai Waziris that inhabited the Wana Plain; their reputation as fearsome warriors was legendary.

The journey from Mir Ali to Wana and then on to Laman had proved uneventful apart from the occasional Pakistani army patrol they saw once they had crossed into South Waziristan. On one occasion a larger convoy of military vehicles had passed them by but the occupants seemed preoccupied, looking ahead. They were clearly not on the lookout for terrorists. Roadblocks and checkpoints were noticeable by their absence. Perhaps the Pakistani army was now gearing up for its long-awaited push into North Waziristan.

En route they had only stopped for petrol once and had made good time. Throughout the trip Kashmiri and his colleagues obeyed the now well-established rules about the use of mobile phones. As the crow flies, the journey from Mir Ali to Laman was close to 90 kilometres. In practice of course the actual route had to follow the limited roads that provide a reasonably safe route to cross the imaginary border into South Waziristan. Taking any other route would have invited trouble as a Toyota Land Cruiser is not a standard sight in the high mountains that bestride the border with Afghanistan. They would, quite literally, have stood out a mile.

On arrival in Laman the plan was to meet up with some of their colleagues in the shade of an apple orchard on the outskirts of town. Their arrival coincided with that of several of their friends. Once the pleasantries had been completed, the fourth most-wanted man in Pakistan sat down with his eight friends to discuss what they were going to do next to avenge the death of Osama bin Laden. There was work to be done. But first tea would be served.

Unbeknown to them, orbiting just out of audible range, a drone was watching their every move. The drone circled as its operator, based thousands of miles away in another deserted area of the planet, fingered the launch switch. He was the latest in a string of people who had watched Kashmiri make his journey. An attack on the vehicle had been considered as it travelled through the remote areas of Pakistan but it was felt it was better to wait and see where he might be going. For Ilyas Kashmiri and his friends there would be no more sunsets over the hills of Pakistan: they were about to become yet another statistic in the global war on terror. He would become the latest victim of a technology that had been developed over nearly 100 preceding years.

CHAPTER 2

Early Days

Ab initio

Throughout their first century of development for military purposes UMA have been focused on a relatively small number of missions. These involved spotting the enemy (surveillance and reconnaissance), reducing the effectiveness of an adversary's defences (suppression), carrying a warhead to a specified location (ground attack/strike) or acting as a target for gunnery or missile practice (target drone). Their ability to carry out these missions has depended upon technological advances in areas such as propulsion, power management and radio and sensor systems.

Reconnaissance has always been recognized as a key element of military operations. Chinese military strategist Sun Tzu (author of *The Art of War*) saw it as an essential element of the preparation for conflict. His teachings emphasized the need to understand the enemy. Before the ideas of powered flight were fully developed, balloons had already showed the value that airborne reconnaissance could bring to the battlefield. In the American Civil War the Union army employed five balloons which were manned by people who formed the first 'aeronaut corps'. They were used solely for reconnaissance purposes, although even this could be a hazardous mission.

Fortunately for the pilots involved, the American Civil War did not involve the kind of immovable trench warfare that was to occur just over fifty years later in France. Their missions did not involve flying over enemy lines and having to land and escape in order to report what they had seen.

The demarcation between friend and foe in the First World War was straightforward. For the balloonists operating in the American Civil War the task was to try to locate movement of Confederate troops on what was quite a dynamic battlefield. The focus on reconnaissance from the air also continued into the First World War. These were, of necessity, manned operations. That remained true, even when the simple camera shots of the battlefield were taken from balloons over the Somme and Ypres.

In the interwar years the exhaustion of the First World War allied with a national economic crisis reduced investment in wartime technologies. The

First World War was supposed to be the last. That hope, however, started to evaporate in the 1930s as Nazi Germany began to rearm itself and issue bellicose statements about its various long-standing grievances over places like the Sudetenland.

Throughout this period, research into technologies that might have a military application was encouraged, although that does not mean it was given *carte blanche*. The classic corrective levers such as jealousy and interservice rivalries ensured that the potential of some ideas was not immediately obvious. The problem of the psychology of 'not invented here' applied then as in many ways it still does today.

When early designs were developed for the concept of a pilotless plane the Luftwaffe were not impressed. They, after all, had the Spanish Civil War as a proving ground for their new array of aircraft and the results seemed to support that they had spent their limited development funds wisely. It was only during the Battle of Britain that the weaknesses in the order of battle of the Luftwaffe became clearer. As the Allied bombing pressure built up on Germany, so they needed a means of retaliation. It was at this point that the profile of the pilotless aircraft that was to become known as the V-1 emerged. It was a weapon aimed at exacting a toll in retaliation for the day-and-night raids by Allied bombers deep into the heart of Germany.

The V-1 was not the first incarnation of a flying bomb. That had come in the very early part of the twentieth century with the development of the aerial torpedo. This was a simple concept that saw an unmanned aircraft carry a warhead through the air to its target. Actually building a reliable and robust solution to that requirement was not quite so easy.

Years later in 1977 when Abe Karem arrived in the United States he was interested in a slightly different problem. He is widely regarded as the person who rescued the UMA programme in the United States. When he arrived it was in a dire state. Borrowing $18,000 a year from his family, he built up a business building UMA. One of his first was the Albatross. The Amber came later and had an endurance of thirty-eight hours. Karem recognized the value of having large wings, hence the name Albatross.

In an interview with the *Economist* in December 2012 he was keen to distance himself from being the man who armed the Predator UMA. He said he simply wanted UMA to 'perform to the same standards of safety, reliability and performance as manned aircraft'. At the time, the most promising UMA available in the United States was the Aquila but it took thirty men to launch it. Even the V-1 did not require that kind of manual support. It also had a poor track record from a reliability perspective. Reports appearing at the time suggested that one crashed on average every twenty hours. It seemed that the idea of creating UMA had outpaced the existing technology. Karem set out to directly change those perceptions with his own developments.

In time Abe Karem's methodical approach to the development of UMA would create the basis from which the Predator and Reaper would evolve. With the help of the United States Defense Advanced Research Projects Agency (DARPA) the various important technological breakthroughs in flight control systems, data links and sensor systems emerged. The decision to arm UMAs, however, was not something he had envisaged.

As Karem was keen to point out, he was not the person who armed the platforms he had been instrumental in developing. That came about because of the juxtaposition of certain military and political necessities. After the end of the Cold War, conflict took on new characteristics. One of those involved hunting for people allied with international groups that chose to hide in states with little in the way of an effective security apparatus to pursue such individuals and groups.

A new form of undeclared war started. Instead of an act of war being defined by a manned intrusion across a border into another state, military operations would be conducted by unmanned aircraft. They would track people who would plan acts of terrorism across the world into the shadowy locations they chose to make their temporary homes and kill them. The message of armed UMA was very simple. Nowhere was safe.

Once the decision to arm them had been made, it was only a matter of time before other countries would follow suit. As the twenty-first century enters its second decade the proliferation of armed UMA is one of the more obvious developments in the order of battle of many of the world's air forces. It was an inevitable outcome on a journey that can be traced back nearly 150 years.

Unmanned air power

It was not the first time someone had thought of the idea of using unmanned platforms to deliver some form of military effect. On 22 August 1849 the Austrians launched over 200 pilotless balloons carrying bombs controlled by timing fuses over the city of Venice. When the balloon exploded, the bomb would drop on the city. The scheme had been drawn up after the Austrians decided that it was difficult to move artillery close to the city because of the difficulty of traversing the lagoons.

An article published at the time in *Scientific American* suggested that up to twenty-five bombs a day might be delivered into the city. Each balloon was 23 feet in diameter and carried 33lb of explosives. One actually detonated over St Mark's Square. The obvious flaw in this scheme, however, was all too apparent when the wind changed direction. What was needed was some way to ensure that the means of delivering the bombs could be controlled. Balloons in their original unpowered configuration were simply not the right platform.

It was at around this time that Sir George Cayley resumed his interest in aeronautics after a gap lasting nearly forty years. His first interest in the subject had been piqued at the end of the eighteenth century. At the age of 26 he produced the first detailed analysis of the founding principles of aviation: weight, lift, drag and thrust. He sketched out an idea for a fixed-wing aeroplane. He saw the problem of flight as one of overcoming drag through generating lift using a source of propulsion and then having controls that could stabilize the aircraft in flight and ensure it could turn.

The culmination of this saw him publish a series of three papers called 'On Aerial Navigation' that were published in 1810. In this he revealed that he had built and flown a man-size glider at Brompton in 1804. His experiments in aeronautics led him to discover the benefits of a cambered aerofoil. He is also credited with the discovery of the importance of the dihedral angle and its impact upon stability in flight of an aircraft around the roll moment. His next glider appeared in 1849 and was flown over a distance of several hundred yards at the same location where he had flown nearly fifty years earlier. Sadly, as his design work was coming to fruition he died in 1857 at the age of 83.

In 1862 and 1863 two patents received approval in the United States. The first involved the development of a flying machine that could carry a bomb. The second, which was awarded to Charles Perley of New York, concerned the design of an unmanned bomber. This was nearly forty years before the Wright brothers finally took to the air.

One pioneer of this period was Samuel Pierpont Langley. He was one of a number of aviators that started their work towards manned flight by building unmanned gliders. Langley began his experiments with gliders in 1887. Two years later he achieved his first success when his Number 5 unpiloted model flew for nearly three-quarters of a mile, having been launched from a catapult on a boat on the Potomac River. The distance he achieved with this experiment was nearly ten times that previously accomplished by a heavier-than-air flying machine.

For Langley it was the first step on a glittering career that would eventually see his name immortalized when the United States named its preeminent flight research centre in honour of his work. Langley, however, was only using the development of the powered glider to create a manned flying machine. He had no interest in developing a UMA.

Given the remarkable coverage and associated rapid development of flying machines that then occurred, it is remarkable that the whole idea of unmanned aircraft managed to continue to germinate. At the time the development was not led by specific military requirements. This made selling the potential of unmanned aircraft to the military more difficult. Those involved were developing a technological capability that some doubted had any practical value from a military perspective.

Nearly 100 years later this same idea was being explored by organized crime groups in Italy as a weapon of assassination. In this case the platform was a model aircraft and its controller a simple hand-held device equipped with two joysticks to allow the operator to adjust the power levels derived from the engine and alert the control surfaces on the model. This simplistic need to somehow effect control over an unmanned aircraft was one of the early challenges facing designers.

However, in the early part of the twentieth century two important developments had to occur before the flying torpedo could finally be tested in anger. The first of these was the ability to transmit signals using radio waves. The second was to be able to use a gyroscope to control the attitude of the platform being used to carry the warhead. Controlling altitude was not such a difficult problem given advances in the design of barometers.

History is opaque when it comes to deciding who was the first person to realize that radio waves could be transmitted over significant distances. Nikola Tesla was one of a small number of researchers active in the field at the time. His first tentative steps into aviation were taken in 1877 when he attended the Polytechnic School in Graz, Styria. His father had selected the institution for its reputation.

It was here that Tesla's fertile mind began to flourish, excelling in his exam results. He was encouraged by several of the academics at the college to explore his ideas. Tesla forged a close relationship with Dr Alle who instructed him on calculus. His own specialization was differential equations. In one discussion between them Tesla outlined the idea for a flying machine that he had conceived based upon what Tesla himself in his own words called 'sound scientific principles'.

At this point towards the end of the nineteenth century Tesla was not alone in thinking about how to develop a flying machine. But he was one of the few to underpin his work in such a way. Many others in the field at the time adopted approaches that were based on what might today be described as an incremental build approach: accepting that intermediate designs would not necessarily work.

Guglielmo Marconi was another man working at the forefront of scientific understanding. Around the same time both men had been actively engaged in looking at the ability of radio waves to carry signals over long distances. While their first motivation was to develop systems that could replace the telegraphy systems that had sprung up in places like America, other applications would emerge. Each man, however, would take a very different approach to the problem.

Tesla's work was initially confined to quite low frequencies. In 1899 he flabbergasted an audience on a wet, rainy day at Madison Square Garden when he demonstrated the ability to remotely control a small boat sailing on a demonstration tank. On that day he also managed to show that a similar degree of control could be achieved over a small demonstration model sailing under the water.

This simple – but for the crowd amazing – demonstration was the precursor to both unmanned aircraft and unmanned underwater vessels. In a series of articles in the journal *Electrical Engineer* Tesla had trailed the ideas that had led him to conduct this impressive demonstration.

At the heart of Tesla's invention was a device that became synonymous with the development of the computer. It was a simple logic gate that enabled switches and controls to be toggled depending upon the reception of different signals. The device forms the logical basis of what today is known as the AND gate. This recognizes when a signal is present at both of its inputs and provides an output signal. For all other combinations of signals the output remains unchanged.

Using this mechanism Tesla was able to demonstrate the simple idea of steering a model to the right or to the left. While for the majority of the crowd this was a stunning achievement – many thought he was actually using thought control to change the heading of the model – Tesla himself saw this as only the start of a much wider range of applications of his fundamental research into radio systems.

In practice, of course, Tesla was using the very same ideas that are used today by amateur radio-control enthusiasts to control model airplanes, ships and motorized vehicles. Each control was hard-wired to a specific frequency. By adjusting the levels on the control he created signals that would drive servo-systems and adjust the controls to create the effect of remotely steering the vessel.

A few months earlier Tesla had been granted a United States patent covering his invention. It was numbered 613809 and entitled 'Method of and Apparatus for Controlling Mechanism of Moving Vessels or Vehicles'. It covered 'any type of vessel or vehicle which is capable of being propelled and directed, such as a boat, a balloon or a carriage.' His more general name for the technology was teleautomaton. In the lineage of remotely-controlled platforms (the more generic phrase used today), it stands out as one of the essential pillars on which future developments would rely.

Marconi's radio trials
Ask the man in the street who invented radio and it is likely that many will name Marconi. Over the last century his name had become closely associated with the initial developments in radio systems. Marconi, however, was investigating a more promising part of what was to become known as the electro-magnetic spectrum. His work proved to have greater potential when it came to transmitting signals over longer distances, a prerequisite to today's unmanned drone operations in places like Afghanistan.

Establishing a precise date when Marconi was able to transmit and receive messages is difficult but in the summer of 1895 he had taken his equipment outside and was able to achieve the reception of signals over a distance of 2.4 kilometres.

Unable to attract investment in Italy, he moved to England where his work was quickly recognized by the Post Office. Its Chief Engineer William Preece recognized the potential in the 21-year-old's work. Preece had studied under Michael Faraday at the Royal Institution in London and had been responsible for introducing a number of new signalling systems into the burgeoning railway network that was springing up across the United Kingdom. In 1889 Preece had been able to transmit and receive Morse code radio signals over a range of 1 kilometre at a test site on the edge of Coniston Water in the Lake District in England.

Under Preece's tutelage Marconi was able to continue the development of his work. Preece secured funding for Marconi from the Post Office. In March 1897 Marconi had been able to increase the distance over which he could transmit radio signals to 6 kilometres. Barely two months had passed before Marconi transmitted a signal across the Bristol Channel, the first time such a transmission took place over the sea. Within days that had increased to 16 kilometres.

In 1899 the first radio signal was broadcast across the English Channel. Perhaps the major breakthrough came when Marconi installed equipment on the American Line's SS *St. Paul*. On 15 November 1899 the ship became the first to report her imminent arrival using wireless. The message had travelled a distance of 122 kilometres (66 nautical miles) from the vessel to the receiving station set up on the Needles on the Isle of Wight.

It was not long before Marconi was setting out to try to send radio signals across the Atlantic. His first attempts, however, were not as successful as some historical accounts suggest. Marconi was unaware of the impact of the ionosphere in refracting radio waves. What many people believe to have been the first successful transmission of radio waves across the Atlantic was, in fact, a failure.

Stung by the sceptics, Marconi installed equipment on the SS *Philadelphia*. As it sailed west from the United Kingdom Marconi recorded signals sent from the Poldhu station in Cornwall. What Marconi quickly appreciated was that the best ranges were achieved at nighttime. This was when the wavelength for the transmissions Marconi was working on propagated better at night.

In 1902 the first verified transmission took place across the Atlantic. Within two years a commercial service had been established that would transmit daily news summaries to subscribing ships. Ten years after the first transmission across the Atlantic Ocean the same receiving station in Massachusetts would be the first to receive the distress signals originating from the stricken RMS *Titanic*. Marconi's research had led to communications systems that could send simple telegraphy messages across the expanses of the Atlantic Ocean. Having created the basis of radio communications, the question was to what other applications might it be applied? One area to emerge was the idea of remotely controlling an aircraft.

To remotely control an aircraft a remote pilot must be able to communicate signals that adjust the main flight controls (rudder, elevators) and engine power on the platform. Contemporary radio-controlled model aircraft work on the same principles defined at the outset of the development of UMA back at the start of the twentieth century.

The operator moves a joystick to input a change in the settings of one of the flight controls. A signal whose amplitude is directly proportionate to the degree of movement is generated. This then modulates a carrier signal which is subsequently transmitted to the receiver in the aircraft.

The approach to modulating the carrier wave can be simply using its amplitude (Amplitude Modulation: AM) or by inducing a change in frequency that is proportionate to the level of the signal (Frequency Modulation: FM). Each of the two approaches had some advantages and disadvantages. The earliest models used AM as the approach as FM was not invented until the 1930s.

In the receiver the original signal is recovered through a process of demodulation and used to drive servo controls that change the settings of the flight controls. Typically four separate communications channels are used to send the signals for adjusting the main flight controls. For more advanced situations, such as those involving the need to raise and lower an undercarriage, additional channels would be needed.

Initial flight trials

The other major hurdle to creating an unmanned aircraft was the issue of how it would be controlled. This was a subject that an American inventor called Elmer Sperry had turned his mind towards. Sperry was a serial inventor. By the time of his death in June 1930 he had over 400 patents lodged with the United States government.

Of all these, perhaps his development of the gyroscope was the greatest. This allowed control mechanisms to be created that could stabilize ships and aeroplanes. It also provided the wherewithal for pilots to have a reliable artificial horizon; an essential element enabling them to fly in fog. Just before Sperry's death a United States army plane equipped with two of his gyroscopes flew without any manual intervention on a south-westerly course the 140-kilometre trip from Sacramento to San Francisco.

For Sperry it was the culmination of work that he had first started in 1896, around the same time that Marconi was taking his first tentative steps in radio. Over the next fifteen years Sperry would build gyrocompasses for battleships and equip destroyers with gyrostabilizers. By 1911, however, Sperry had turned his mind away from the naval domain into the aviation world.

The prospect of remotely-controlled aircraft intrigued him. He understood how the radio signals could be used to control the plane. That was not too difficult. Actuators could be commanded to move control surfaces to manoeuvre the aircraft. Within the decade, many of the control surface manipulation problems would have been resolved and unmanned aircraft would move from the drawing board into actual flying machines.

However, as Sperry saw it the problem was how to stabilize the platform once any change had taken place. He could see that taking off was not a key problem. The real challenge lay in maintaining stable flight conditions once the unmanned aircraft was in the air. Too many control inputs would make it difficult to control the platform. Something had to automatically return the platform to a stable configuration once a control input had been made. If an unmanned aircraft was to be feasible, at the very least the remote pilot would have to be able to compensate for wind effects. If that could be overcome then the potential to build a remotely-controlled flying bomb that could deliver a payload against a remote target would really exist.

Unlike the initial configurations of the Predator and Reaper systems, the first generation of unmanned aircraft were seen to be something akin to a flying torpedo. This was the ultimate precursor to the kind of cruise missiles that now fly over 1,500 kilometres to their targets using terrain navigation techniques to ensure they achieve an increasingly precise effect. Given this direction of travel from a development viewpoint it is easy to understand how naval applications became a focus for the application of unmanned aircraft.

Unmanned aircraft were not originally seen as potential intelligence-gathering platforms. They were to be launched on a one-way mission to a target. The idea of a flying torpedo emerged because the initial funding came from the United States navy. The potential benefits were that an aircraft configured in this way could attack an enemy warship over a longer range and deliver the payload more accurately than a traditional torpedo riding through the water.

The main problem with the concept of remote control would be where the operator would be based. At sea level line of sight restrictions would ensure that any warship being attacked over the horizon would be relatively immune. The observer/pilot would therefore have to fly somewhere close by to ensure they could maintain line of sight on the aircraft.

For those unconvinced by the arguments over the utility of unmanned aircraft, this was nonsense. If you needed to fly the controlling aircraft near to the unmanned platform it would be vulnerable to counter-air fire from any target vessel. Lose the controlling aircraft and the unmanned escort would quickly become ineffective. The same arguments were also difficult to make at a time when manned flight was seen to be at the leading edge of technological development. It seemed as if every day in the civilian field new records and achievements were creating headlines in the world's press. Man had finally overcome gravity. Why now find ways of taking that away from him and handing that task over to machines that were capable of being operated remotely?

The idea of using an escort aircraft to overcome the problems of the radio horizon was regarded at the time as simply not practical. Unaware of some of the limitations on the operation of the existing technology, a plan to use the flying torpedo to attack the strategically vital submarine pens at Wilhelmshaven in Germany was proposed towards the end of the First World

War. While that attack was eventually abandoned, twenty-five years later a modified B-24 aircraft was used to conduct an unmanned attack upon the German submarine base at Heligoland.

The equipment used aboard the aircraft comprised a gyroscopic stabilizer, a directive gyroscope, an aneroid barometer and servo-controls to adjust the ailerons and rudder. Additionally a device that measured the distance travelled was also installed. Initial trials proved that the accuracy achieved meant it could not be used to hit a moving ship.

However, the idea was felt to be maturing to the point where the army may have an interest in using the weapon in a tactical role. Its range of between 80 to 160 kilometres was felt ideal for a battlefield role. As far as the United States navy was concerned, the concept was interesting but the technology simply could not deliver an effective solution. The technology of unmanned aircraft had not yet developed far enough for them to play any significant role on the battlefields of the First World War.

Once America had entered the war, Sperry tried again to interest the navy. After some debate the Secretary of the Navy allocated $50,000 for the development of an initial capability. Five Curtiss N-9 seaplanes were initially allocated. Two more were to be added to the programme over time. In the event the Secretary of the Navy was so enthused by the concept that the initial investment was quadrupled.

The first flights of the aircraft took place in September 1917. A pilot conducted the take-off before handing over the aircraft to the automatic pilot. Two months later on a test flight the autopilot successfully controlled the test aircraft as it flew 50 kilometres to a target before dropping a sandbag as a simulated weapon. The attack had an accuracy of around 3 kilometres. Five years later flying on another aircraft the same capability was demonstrated against targets out to 140 kilometres.

Around the time of the first test flights Rear Admiral Ralph Searl, the Chief of the Bureau of Ordnance, wrote up some ideas on how to win the First World War quickly. He identified the German U-boat submarine bases at Wilhelmshaven, Cuxhaven and Heligoland as being priority targets. Admiral Searl suggested this could be a mission for the flying torpedo. The idea was to move a vessel carrying the unmanned aircraft close to the German coastline before launching the flying torpedoes. Despite the suggestion, the Chief of Naval Operations was not impressed. He had concerns over the range and accuracy of the weapon.

A demonstration was, however, organized for the Chief Signal Officer of the United States army, Major General George O. Squier. He did see some potential applications in the land environment and initiated a separate project that established a base at McCook Field in Dayton, Ohio. A clear emphasis was placed upon ensuring the costs were minimized and that the platform could be produced in large numbers. As a result, little development occurred on the airframe.

However, as a direct result of the tests on the N-9 in September 1917 an order for six special planes was awarded to the Curtiss Aeroplane and Motor Company. This was a departure from the previous tests using the N-9. The airframes involved were to have an empty weight of no more than 500lb (227 kilos) and they were to be able to fly a distance of 50 miles at a top speed of 90 miles per hour. Their payload was to be 1,000lb (454 kilos) of explosives. Importantly, the aircraft were to be fitted with the facility to allow them to be controlled remotely.

Within the thirty days stipulated in the contract, the first of the planes was handed over for testing. The platform had never been flown as a piloted aircraft, let alone in an unmanned configuration. Problems quickly emerged. The aircraft was incapable of taking off under remote control. It soon became clear that the aircraft's flying characteristics had to be carefully assessed.

Lawrence Sperry, the son of Elmer Sperry, decided that he would adapt the aircraft so that he could manually fly the take-off before handing the platform over to the autopilot. His attempts were partially successful. On the first flight he wrote off the aircraft when it hit a divot while taxiing. On the second attempt he managed to get the aircraft into the air. However, when he engaged the autopilot the aircraft performed what was described as two lateral flips before he regained control.

It was clear that this was insufficient to resolve handling problems with the airframe. Showing some ingenuity Lawrence Sperry devised a means of attaching the aircraft to a motor car. They then drove it along the Long Island Motor Parkway at 80 miles an hour. It was a unique form of wind-tunnel testing: on the move. This allowed them to make subtle adjustments to the control surfaces and the autopilot. Two further attempted automatic launches saw one successful flight and a failure. Two further successful flights were to take place where the aircraft managed to fly 100 yards before crashing into the ground.

The aircraft selected to be the platform for the second iteration of the American experiments was the Verville-Sperry Messenger. Its original mission had been to provide a form of airborne battlefield motorcycle, carrying messages to commanders in the field. It was a rugged platform that was easily able to land in small fields. This and its simple control surfaces made it an ideal candidate to be converted into an unmanned aircraft. In wind-tunnel tests the aircraft was shown to have good longitudinal stability. This characteristic was also present at both high and low speeds.

The aircraft was also reported to be easy to control and did not react too quickly when commands were input to the control surfaces. This was an important factor for any form of remote control as too short a time lag between control input and reaction could have created instability if the operator tended to over-control the aircraft.

The flight handling characteristics were very similar to the Sopwith Camel SE.5, while its manoeuvrability was compared favourably with the versatile Nieuport aircraft developed in France. In total twelve were converted into radio-controlled flying bombs or torpedoes. They were known in the Army Air Service as the Messenger Aerial Torpedo (MAT). None of these machines was ever used in anger.

Another variant of an unmanned aircraft that appeared during the First World War was the Kettering Bug. It arose from the requirement to strike at enemy targets at ranges of up to 75 miles while travelling at 50 mph. With America now involved in the war, the United States army had originally asked Charles Kettering to design an unmanned aircraft (or flying bomb) that could fly to a target over a distance of 40 miles. Initial flight tests with the first variant of the aircraft were not encouraging. Racing down a small track, reminiscent of the means used by the Wright brothers to launch their first manned aircraft, the Kettering Bug simply toppled forward and collapsed in a heap when it reached the end of the take-off run.

Eventually the problems were ironed out and a small number of Kettering Bugs flew. Archive footage taken at the time from a chase aircraft shows the Kettering Bug apparently 'twitching' in flight. This was due to inputs from a simple gyroscopic stabilizer that was guiding the aircraft to the target. On board a small counter linked to the rotation of the engine calculated when the aircraft had travelled the requisite distance to the target. At this point the engine was shut down and the aircraft entered a ballistic trajectory to the target. On impact the payload of 180lb (82 kilos) of high explosive detonated. This was one-third of the total weight of the aircraft.

Although the reliability of the Kettering Bug improved, the aircraft was never deployed in anger in the First World War. By the end of the war forty-five of the airframes had been assembled. When funding for the programme ran out in the 1920s the design information, which was secret, was consigned to the archives. As far as the United States was concerned the latter part of the 1920s and the early years of the 1930s were a time of little interest in unmanned aircraft. They lacked the sense of an immediate operational requirement for this kind of military capability. Limited funds were available anyway and any ongoing developments were funded out of research grants. It was, however, to re-emerge in 1935 due to a meeting that took place in London.

The Chief of Operations of the United States navy, Admiral Standley, was in the capital to attend the 1935 London Disarmament Conference. This started in December and had some ambitious goals to try to limit the size of naval units. At the meeting in a sign of things to come Japan vetoed an attempt to place restrictions on the numbers of warships that any country could operate. Italy also failed to sign the final treaty. However, France, the United Kingdom and the United States did sign. With the dark clouds of war already gathering over Europe, the treaty was not to remain in force for very long.

Once the meeting was over Admiral Standley went on to hold some bilateral discussions with his opposite number in the Royal Navy. It was during the course of these conversations that the admiral learned of British trials with remotely-piloted aircraft that had recently culminated in the development of a new target drone. This was called the De Havilland Queen Bee.

Admiral Standley, who had a distinguished naval career already spanning forty years, was enthusiastic about the British developments. With Germany re-arming and Japan failing to sign the disarmament treaty it was just possible that another war was around the corner. He left London clearly determined to look again at the potential role unmanned aircraft might play in the United States navy.

Early unmanned aircraft research in the United Kingdom

British interest in the use of unmanned aircraft had a slightly different starting point. How to deal with the Zeppelin raids? At the Royal Aircraft Establishment (RAE) at Farnborough in England research efforts were under way into how to fly an unmanned aircraft.

Zeppelin raids over London and the south-east of England were a huge problem. They were affecting the morale of the people. What was needed urgently was a means of attacking the Zeppelins that would result in the airships being shot down. At the time British air defences, such as anti-aircraft fire, were not that effective. At first they were divided between the Royal Navy and the British army. In February 1916 the British army took full control. Some guns were converted to an anti-aircraft role with 271 being installed by the middle of 1916 alongside 258 searchlights.

DRONE WARFARE

The air defence element of the defence of the United Kingdom was also fragmentary. It was divided between the Royal Flying Corps and the Royal Naval Air Service (RNAS). The latter took responsibility for engaging the Zeppelins before they crossed the coast over the North Sea. The RFC then took on the task when the Zeppelin had crossed the coastline. In February 1916 fighter strength for the defence of the south-east of England was just ten squadrons and many of these were underequipped.

The main issue, however, was the armament on the fighter. The Vickers-Challenger interrupter mechanism that allowed bullets to be fired through the propeller was still several months away from becoming operational. Experimentation had proved it could work but it was not yet ready to be fitted to the B.E.12 fighters trying to bring down the Zeppelins. Initial trials with incendiary bullets were also unimpressive. There was also some uncertainty over the structure of the air bags on the Zeppelin. This led to suggestions that the airships were fitted with an outer envelope of inert gas to avoid ignition by incendiary bullets. Other more innovative approaches were therefore needed.

While it is difficult to be certain how many Zeppelins were destroyed by British air defences in the First World War, at least three incidents are documented. The first Zeppelin was brought down over Ghent in Belgium on 7 June 1915 by Sub-Lieutenant Reggie Warneford from the Royal Naval Air Service. During his flying training Warneford developed a reputation for aggressive flying. His first encounter with a Zeppelin had been less successful. On 17 May 1915 he tried to bring down Zeppelin LZ.39 as it approached the United Kingdom. Despite using a machine gun loaded with incendiary ammunition he failed to destroy the target. The airship simply ascended out of range by jettisoning ballast.

Days later Warneford brought down Zeppelin LZ.37, dropping six 20lb incendiary devices on the airship from above. It was a very brave and novel attack delivered with his customary panache in the face of a barrage of defensive fire from the Zeppelin. The last bomb succeeded in setting the target on fire. Of the crew of Zeppelin LZ.37 only one man survived, the helmsman, after it crashed in Sint-Amandsberg in Belgium.

The drama of the engagement was not over yet. The updraft from the explosion caught Warneford by surprise and flipped his Morane-Saulnier Type L aircraft onto its back in the air. His engine also cut out. Showing incredible calm in what was a very difficult situation, Warneford regained control of his aircraft and glided to land behind enemy lines. Repairs to the engine took him thirty-five minutes before he re-started it and took off to return to his base. For this Warneford was awarded the Victoria Cross. Sadly, he was to die only days later when the aircraft he was flying suffered a major structural collapse.

Just over a year later on 24 September 1916 Second Lieutenant Sowrey of 39 Squadron shot down the German navy Zeppelin L.32 over Great Burstead. In a more famous incident on 2 October 1916 Second Lieutenant Tempest of 39 Squadron shot down German navy Zeppelin L.31 over Potters Bar. Zeppelin L.48 was also destroyed by a B.E.12 on 17 June 1917. The fighter aircraft deployed at the time were simply inadequate. As the Zeppelins improved and were able to operate at higher altitude they simply became out of reach of the fighters which had a ceiling of 12,500 feet. Their rate of climb was also poor, taking eleven minutes to reach 5,000 feet.

But the idea of flying an unmanned aircraft carrying a warhead into a Zeppelin offered an alternative solution. The first generation of such a device was constructed at the P. Hare Royal Aircraft Factory in Putnam. The idea had come from Captain Archibald M. Low of the Royal Flying Corps signals unit at Feltham. It was called the AT: aerial torpedo. Years later at the end of the Second World War a manned version of this same idea was to appear over the Pacific Ocean. This was the fabled kamikaze.

The design of the AT was quite simple. It was a shoulder-wing monoplane driven by a two-cylinder ABC air-cooled engine that was able to produce 35hp. The radio antenna designed to allow it to be remotely controlled was affixed down the side of the fuselage. The overall weight of the aircraft was 500lb (227 kilos).

To achieve lateral control the wings were bent (warped) and stability was achieved by them being shaped at a dihedral angle. Six test aircraft were built. Its first flight occurred on 6 July 1917. The aircraft took off almost vertically, entered a stall and crashed. All of this occurred before the radio could have any effect on the controls. The second aircraft never left the ground, simply running along until its undercarriage collapsed. The third test also came to a quick end when the engine failed shortly after take-off.

In 1922 the RAE started testing its RAE 1921 Target aircraft. The results were not encouraging. All of the test aircraft flown from an aircraft carrier simply crashed into the sea. Controlling the aircraft at low speeds was clearly a problem. To resolve this issue a small radio system was added to provide control inputs from the point of take-off. What had up until then been a string of failures was halted. In 1924 the RAE Target 1921 flew for thirty-nine minutes at speeds of up to 100 mph. It flew for a distance of 65 miles.

The second generation of the design was quickly forthcoming. A monoplane called the Larynx was designed that could operate over a range of 100 miles. Its name was derived from a highly-contrived acronym that read Long Range Gun with Lynx Engine (LARYNX). Work on it started in 1925. It was capable of flying at a speed of 200 mph. This was surprisingly quick for the period and showed what was possible when the weight of a pilot was removed from a flying machine.

Its first test flight took place on 20 July 1927 in the Bristol Channel. It was launched from the S-class destroyer HMS *Stronghold* located in Swansea Bay. The aim was for the vehicle to fly to a point around 10 nautical miles north of Cape Cornwall. This was a distance of 200 kilometres (108 nautical miles). At the end point of the flight a drifter was anchored to observe the final moments before the aircraft hit the water. To help the observers at the target point pick up the aircraft, titanium tetrachloride was to be ejected from the platform in the last 5 miles of the flight.

The outcome of the flight was to be somewhat disappointing. When the engine was opened to full throttle a junior member of the team from Farnborough was due to make some final adjustments before it was released. At this point the trolley carrying the Larynx collapsed and the aircraft crashed forward onto the catapult causing the propeller to disintegrate. To compound matters the container carrying the titanium tetrachloride burst open and the unwitting junior scientist from the RAE was projected by the tailplane over the edge of a packing case onto the steel floor of the destroyer. It was an ignominious start to a career that would eventually see Dr Gardiner appointed as Director of RAE in 1955.

Other tests, however, produced more positive results. Five of the aircraft were then sent out to the RAF airbase at Basrah in Iraq. Testing was to involve live warheads each weighing 250lb (113 kilos). The first four tests again were inconclusive before the fifth crashed, having flown successfully in May 1929. Arguably this was the first cruise missile to fly over Iraq. It set a precedent that was to be repeated sixty-two years later at the outbreak of the First Gulf War.

Building on the developments made during the Larynx programme, the Royal Navy was anxious to develop a new series of target drones to help train naval gunnery teams. They needed a target that could manoeuvre to simulate the kind of airborne attack that might now occur on warships. The result of this saw the development of the De Havilland Queen Bee.

On his return to the United States Admiral Standley asked his research teams to develop a similar capability. In his book *Unmanned Aviation: A Brief History of Unmanned Aerial Vehicles* Laurence Newcome details the admiral's requirements. What he was after was a radio-controlled seaplane that could fly at 100 knots to a ceiling of 10,000 feet.

Importantly, given the developments in dive-bombers, the unmanned aircraft should be capable of not only flying straight and level but also climbing, turning, gliding and descending into a 45° dive before pulling out. Throttle controls were to be fully under remote control by radio out to a range of 10 miles from the host vessel. Take-off was to be conducted either conventionally or using a catapult-assisted mechanism. Within a year the United States had developed its first target drone. Its control surfaces and throttles were manipulated through twelve radio channels.

In March 1937 the target drone and its controlling aircraft flew for the first time. A year later Adolf Hitler sent his German troops to occupy the Sudetenland. War in Europe was now almost inevitable. While in Britain the development of unmanned aircraft was being driven by the need to develop target drones for the Royal Navy, across the North Sea in Germany a far more advanced set of ideas had been traced out on the drawing board. This was the design concept for the world's first cruise missile, the V-1 flying bomb or 'doodlebug'. In June 1944 as the Allied landings started in France it was to make an appearance over the skies of England as the Third Reich tried to bomb the Allies to the negotiating table. It was a plan that would not succeed.

CHAPTER 3

Into War

The evidence shows, beyond doubt of 'planting', that the Germans have for some time been developing a long range rocket at Peenemünde.

Extract from Air Scientific Intelligence Interim Report
26 June 1943

There are probably those of you amongst my listeners tonight who imagine that tackling a robot which cannot shoot back or take evasive action is – to use a popular service term – a piece of cake. Nothing could be further from the truth.

Squadron Leader Joseph Berry
Interview recorded for the BBC Broadcast 31 August 1944

Overview

Aside from the contemporary Predator and Reaper UMA, perhaps the most famous unmanned aircraft is the V-1 flying bomb or 'doodlebug' developed by Nazi Germany. The V-1 carried a 1,870lb (848 kilo) warhead; vastly different to the weight of the explosives carried on the Hellfire missiles carried by today's unmanned aircraft.

Over 9,000 V-1 missiles were launched against England over a period from June 1944 to March 1945. This remains the largest bombardment in the course of war by UMA. This is a launch rate of close to thirty per day, well below the figure of 500 a day demanded by Hitler at the start of the campaign. A total of 2,419 of them landed in London killing 5,126 people (a ratio of roughly two people per V-1 explosion). In the rural areas surrounding London the death toll was significantly lower. In that area 2,789 V-1s killed 350 people. Added together, the overall ratio for the campaign against the United Kingdom was a total of 5,208 V-1s killing 5,476 people. This ratio of roughly one person to die per attack is not dissimilar to similar historical ratios of those killed by iron bombs.

The primary purpose of the V-1 was as a weapon of terror. While the Nazis had high expectations of what the V-1 could achieve, in practice the level of explosives delivered on London was never sufficient to really affect the outcome of the war. It did, however, have a noticeable effect on the British public. Writing in his novel *Unconditional Surrender*, author Evelyn Waugh expressed the *zeitgeist* describing the V-1 as being 'impersonal as a plague, as though the city were infested with enormous venomous insects'.

In the latter stages of the war as the Allied advance across Europe moved closer to Germany the Nazis turned the V-1 on to the port of Antwerp and the cities of Brussels and Liege. The V-1 bombardment of the port of Antwerp was important as it was a major Allied staging post for operations in Europe. The attacks killed 743 military personnel in Antwerp and injured a further 1,078. This is around 50 per cent of the total number killed in England over a longer period. Detailed analysis shows that the death toll per missile fired was the same in Antwerp as in London. It also had similar psychological side effects.

In an article in the *British Medical Journal* published in January 1946 a Captain Freeman explored the effects of the attacks using both the V-1 and the V-2 ballistic missile. Over the 175 days of the attacks he noted 4,248 V-1 attacks on the area of Greater Antwerp. At the same time 1,712 V-2s also landed on the city. Troops based in the area who reported sick during the bombardment by the V-1 and V-2 missiles could be divided into two broad categories. Some showed symptoms of fear in varying degrees and the remainder complained of headaches, weight loss and insomnia. Civilian deaths in Antwerp, however, were higher at 2,900 with 5,433 injured. Other major population centres were also attacked such as Remagen, Arras, Cambrai, Mons and Lille, albeit on a much smaller scale.

When the V-1 arrived over England on 13 June 1944 it created a range of problems for the Royal Air Force and for the air defence systems of the United Kingdom. During the V-1 campaign the Royal Air Force flew around 20,000 interception missions, some of which were quite hazardous.

Pilots employed a range of tactics to bring down the flying bombs. On a small number of occasions a pilot resorted to flying alongside the V-1 and using the wingtip of his aircraft to flip the V-1 over and cause it to dive into the ground. However, this manoeuvre did not always work. One pilot, Flight Sergeant Donald Mackerras from the RAAF flying with 3 Squadron, lost his life while trying to tip a V-1 in his Tempest.

Of course, this had to happen over unpopulated areas if the effect of the warhead exploding was to be minimized. What was also unknown at the time was whether there was a point at which the warhead became armed. Other pilots found that by flying fast and close to the target the wake of their aircraft could create sufficient turbulence to upset the operations of the gyrostabilizing mechanisms built into the missile. If sufficiently high, the V-1 would again dive to the ground.

Pilots who approached with all guns blazing were often surprised when the V-1 ignited, spreading its burning fuel all over the attacking aircraft. It is fair to say some of these engagements were quite precarious. Such was the blinding effect of the explosion that the chasing pilot would often lose vision for up to ten seconds. Fighters were also thrown about as the V-1 disintegrated with some pilots coming out of the engagement upside down. Over time the method of attacking the V-1 was refined and the approach made from astern at a more acute angle.

Records show that Flight Lieutenant Walter from the RCAF flying a 229 Squadron Mark IX Spitfire was the first to be killed on 17 June when the V-1 he was engaging exploded. The epitaph 'killed by a V-1 explosion' was to be written in the records of a total of thirteen pilots during the bombardment. Others died from friendly fire from their own anti-aircraft batteries and one, a pilot from Belgium, is believed to have collided with a V-1. Another flew into the ground while pursuing a V-1 in fog.

Origins

For the V-1 missile many of the design solutions adopted were quite simplistic. As it turned out this was somewhat fortuitous as when it came to be used towards the end of the war the Nazis were experiencing shortages of key materials. Its development had started in the early months of 1937. From the outset the idea was to develop two variants. One would be ground-launched; the other would be flown under the wing of a bomber and be air-launched. This enabled the missiles to achieve a greater geographic coverage in the United Kingdom than would otherwise have been possible solely from ground launches. To confuse any intelligence-gathering efforts by the Allies, the project was given three separate identities. It was variously known as the Kirschkern Programme, F-103 and the V-1.

The team developing the missile was led by Dr Fritz Gosslau. His experience of leading the development of the FZG-43 (*Flakzielgerat*-43) remotely-controlled target drone helped get the development work under way comparatively quickly. Of course at the start of the programme there was little in the way of an operational requirement to define its operational characteristics. This was a technology rather than requirements-led effort. The question for Dr Gosslau and his team was what could be achieved rather than what was operationally necessary.

At one point the design of a remotely-guided bomber capable of carrying a 1-ton bomb load appeared on the drawing tables at the Argus design bureau where Dr Gosslau initially worked. This was called 'Deep Fire' (*Fernfeuer*). The initial reaction of the Luftwaffe to the idea was muted, perhaps surprisingly given the scale of the losses of air-crew during the Battle of Britain. An attempt to get the *Fernfeuer* off the ground failed in the latter part of 1940. Despite the start of the Blitz, the Luftwaffe was not keen on an area bombardment weapon.

The first unmanned missile developed by Gosslau's team was launched on 14 July 1939, a matter of days before the start of the Second World War. In November Gosslau produced a prototype design for a motorized wing-mounted missile that was to be powered by a pulse jet engine. Unlike a normal jet engine, this produces combustion in pulses. For the V-1 this form of power plant offered some advantages in weight and simplicity but it also generated a loud audio signature. This led to it being nicknamed the 'buzz bomb' or 'doodlebug'. However, the engine was underpowered. In flight it could generate 750 pounds force (3,300N) and its static thrust was only 500 pounds force (2,200N). This was insufficient for it to take off unaided.

This necessitated the development of the characteristic ski ramps with which the V-1 design is often associated. Fuel for the missile was not a problem. It could run on any grade, which was important in the latter stages of the war when Nazi fuel supplies started to dry up. The engine was also cheap to manufacture. After all, it was a weapon that was to be used only once. Gosslau moved across to the V-1 design team in the middle of 1942. The first powered flight of the V-1 took place on 10 December 1942 at Peenemünde.

Once launched, it would fly on an established heading. It had enough fuel to fly for approximately half an hour. Its range was set by adjusting a counter that was attached to a small propeller. Every thirty rotations the counter would be decreased by one until it reached zero. A spring mechanism on the rear elevators would then lock them into the down position to drive the V-1 into a near-vertical trajectory. At this point the flying bomb, as it came to be called, would fall vertically onto its target where its warhead would detonate.

The V-1 missiles were manufactured at a range of locations around the Third Reich. The main production facility was located at Nordhausen at the southern tip of the Harz Mountains in northern Germany. This facility gained notoriety through the deaths of nearly 250 people a day who had been forced into slave labour by the Nazis. They were drawn from the nearby Mittelbau-Dora concentration camp which was an offshoot of the Buchenwald camp. It is estimated that in total 20,000 people died from exhaustion, disease and starvation. The regime at the camp was one based on extreme cruelty. If individuals became incapable of working they were transferred to the Majdanek concentration camp in Poland to be murdered.

The manufacturing facility was built underground and was therefore virtually immune to attack from the air. It was used to produce both the V-1 and V-2 missiles. It had been completed in 1943 after Allied bombers had conducted a bombing raid (Operation HYDRA) against the Peenemünde facility on the night of 17 August 1943. Five days later Hitler, recognizing the vulnerability of the facilities at Peenemünde, issued an order to Heinrich Himmler to use concentration camp workers to ensure that production of the V-2 (A-4) rocket proceeded. The initial contract placed with the owners of the facility called for the production of 12,000 V-2 missiles. V-1 assembly started in October 1944. In total 2,275 V-1 missiles were produced at this facility alone.

In total just over 30,000 V-1 missiles were produced between February 1944 and March 1945. The peak rate of production occurred in September (3,419) and October (3,387) 1944. The average rate of production over the period was 2,161 missiles per month. The head of the Luftwaffe Hermann Göring had initially set a goal of 50,000 V-1 missiles being produced per month. This was a wildly optimistic target figure.

As the intelligence picture started to emerge, the Allies dedicated increasing amounts of valuable time to bombing the launch sites. The first raid was conducted in December 1943. Raids by the Royal Air Force on production facilities also had an impact. The Fieseler plant was raided on 22 October, causing the V-1 missile production line to be shut down. This raid was part of a much wider effort by the Allies to disrupt the production of both the V-1 and V-2 missiles. Between August 1943 and March 1945 over 68,913 sorties were flown, dropping over 122,000 tons of bombs targeting the production facilities. This did have an effect in delaying the start of the V-1 campaign. Hopes that it would begin in December 1943 proved unfeasible.

The V-1 did not always carry just a warhead when it flew over England. Some were fitted with a cage that could carry up to twenty-three 1-kilo incendiary devices. On a small number of occasions a facility was added to the V-1 to disperse leaflets and letters from prisoners of war (POWs). This occurred after 28 August 1944. This was a form of what is known today as psychological warfare. Some of the V-1 leaflets contained material showing the impact of Royal Air Force bombing raids in Germany. For Londoners who had endured the Blitz, such propaganda was hardly likely to be received sympathetically. The Royal Air Force had conducted similar operations over

Germany from the very earliest days of the war. As the Luftwaffe was no longer able to conduct raids on London, the aim of using some V-1s to deliver leaflets and messages was clearly to reach a wider population.

The fear was that in some way the use of the letters from POWs was a ruse to try to establish where the V-1s were landing. This was crucial as Allied deception efforts were trying to persuade the Nazis that many of the V-1 missiles landed north of the River Thames. The hope was that the Nazis would reduce the counter on the missile that determined its range, therefore causing it to fall short of the desired targeting area.

The emerging threat

Before the advent of the Predator and Reaper variants of UMA, perhaps the V-1 flying bomb was the most widely-known type of this class of aircraft. It was also known by the designation Fieseler Fi 103. While the Nazi war machine was initially somewhat dubious about the military value of the V-1, as their own fortunes changed so it became increasingly important. It was the only way they could avenge the stream of attacks that were being conducted by the Allies over Germany.

Adolf Hitler placed a great deal of faith in this and its sister weapon the V-2. He believed that such terror weapons could turn the tide of the war, at one point early on in the campaign instructing Albert Speer to focus on production of the V-1 instead of the V-2. In truth the V-1 weapon was more of a last throw of an increasingly desperate regime in Berlin.

German newsreels used air-to-air photography of the V-1 in flight as part of a propaganda campaign. The message was clear: Germany was hitting back at the Allies. The V-1 'secret weapon' was having a major impact on London and the southern counties of England. The language was exaggerated with many false claims of the effectiveness of the missile being made. In the end, however, their belief in the effectiveness of the V-1 was shown to have been misplaced.

The first glimmerings of the development of the V-1 came early on in the war. At a speech in Danzig in Poland when the war was barely three weeks old Hitler referred to 'a weapon which cannot be used against us'. When the words used by Hitler were analyzed carefully it became clear that he was hinting about the existence of some kind of super-weapon. In London, eminent scientists debated what form that weapon might take.

Given the history of the First World War, the initial list of possible candidates inevitably focused on the possibility that new forms of biological or chemical weapons may have been created. Lower down the list reference was made to the possible development of gliding bombs, aerial torpedoes and pilotless aircraft. Other more esoteric weapons, such as a death ray, were quickly removed due to a lack of evidence of their development.

In the early 1940s the Germans had started to develop the V-1 flying bomb. This was used for the first time on 13 June 1944 and then repeatedly until 29 March 1945, a campaign lasting for 289 days. Its use as a terror weapon has some parallels with today's Reaper and Predator missions over Pakistan in terms of its impact upon the civilian population. The V-1 campaign was all about psychological terror. Arguably the Predator and Reaper campaign over Pakistan has created a similar psychological impact upon a wider population.

Despite the developments that had already occurred in the period after the First World War, the arrival into the war of the V-1 and V-2 rockets was initially met with disbelief in some quarters. The intelligence reporting on the development of the devices was fragmented and patchy. Early clues had appeared in March 1937 when a human intelligence source passed over details of a 'small aircraft' that was around 7 metres in length with a wingspan of 3 metres that had a range of around 960 kilometres (600 miles). This, however, was not the only piece of what became a difficult intelligence puzzle to piece together.

The Oslo Report
Intelligence pictures often reply on one important piece of the jigsaw puzzle to help mould disparate and confusing pieces into a more coherent perspective. One element of the emerging picture arrived in London from Oslo. It appropriately became known as the Oslo Report. Reaction to its seven pages of analysis of key Nazi technological developments was mixed. Many saw it as a deliberate attempt to mislead. They cited the variability of the reporting on different subjects. How could one author have such a breadth of knowledge? Others, noting the finer technical details on Nazi developments in radar, saw it through a different lens.

One of the latter was Dr R.V. Jones. He led the effort during the war to analyze what was dubbed 'scientific intelligence'. For Dr Jones the Oslo Report contained information that it would not have made sense to release as part of a deception plan. From his own understanding of the development of radar in the United Kingdom, Dr Jones saw evidence that added weight to the veracity of some of the contents of the Oslo Report. This is important. Intelligence analysis often requires an individual to sift the material carefully.

The problem with the Oslo Report was that the first two notes supplied in it were so obviously wrong. This lent sway to the argument that the report was in fact worthless.

Years later the author of the report was finally named. He was Dr Hans Ferdinand Mayer. At the time of writing the report he was the director of Siemens Research Laboratories. It was a company he was to work for until 1962. In that role Dr Mayer was able to travel overseas. On a visit to Oslo in 1939 he sent an anonymous message to Captain Hector Boyes who at the time was the Naval Attaché at the British Embassy.

In the first message he simply asked if the British were interested in obtaining some of the technical details of advanced Nazi research and development programmes. After a pre-arranged signal had been sent during a routine radio broadcast by the BBC, Dr Mayer spent two days in a hotel in the city typing the seven-page report. It was then mailed to Captain Boyes. That was the easy part. Understanding what was in the report was to prove more difficult.

Some of it was written in what may be described as 'technical-German'. In some areas of development the Nazis were ahead of their counterparts in the United Kingdom. Just trying to analyze what these various terms meant proved to be tricky. This added to the confusion over the accuracy of the report. Some of the material supplied also referenced developments that were available in open sources. However, one sentence in the report was to have far wider implications. It started with the words *Die Erprobungsstelle ist in Peenemünde* (the test site is at Peenemünde).

The picture emerges

That something was happening at the German test range in Peenemünde was clear. The issue was what. One of the problems with analyzing an incomplete jigsaw puzzle is that the analyst has to try to speculate on what might be in the gaps. Trying to fill in the missing pieces when the analyst does not have the front cover is a challenge. That approach is seriously compromised when the analyst is looking for something they have never seen before. Where an adversary has taken a technical leap ahead of your own capability, you have no obvious reference from which to work.

This was the case in the early years of the Second World War. It also caused a great deal of heated debate in the upper echelons of the scientific community. Some individuals simply refused to accept that some of the Nazi developments were feasible. The arguments over the existence of the V-2 rocket, for example, were particularly heated.

As new pieces of the jigsaw puzzle continued to arrive, a conversation overheard between two construction workers provided a vital steer. Further clandestine efforts by a range of spies provided yet more evidence of the existence of a pilotless aircraft. Crucially, sites thought to be launch pads had also been detected in photographic imagery taken over northern France. Within a short period of time ninety-five possible launch sites had been identified. By now those analyzing the information had become convinced they were looking for a small aircraft that could fly from the ski ramps. These were then subjected to a series of major bombing raids by Allied aircraft. It was clear to the Nazis at this point that their secret weapon programme was no longer under wraps.

The launch sites were given the code name 'No Ball'. As the sites were built from reinforced concrete, ground-attack aircraft such as the Typhoon were unlikely to succeed in destroying buildings or ramps. While they could achieve the precision that traditional bombers lacked, the air defences around the sites made any attack extremely hazardous. In the end both the United States and Royal Air Force bomber crews who carried out attacks against the 'No Ball' sites were to pay a heavy price for their efforts to prevent the V-1 attacks from starting.

Long-range photographic missions conducted by specially-converted Spitfire aircraft were to produce slightly more conclusive evidence. After one mission a photographic interpreter called Constance Babington Smith found an object that she had never seen before. She had been searching through a specific set of photographic data that she recalled as being of exceptional quality. Later she recalled that the 'absurd little object was not on the airfield, but sitting in a corner of a small enclosure some way behind the hangars.' She named it 'Peenemünde 20'. It was the first time a V-1 had been seen by the Allies.

This was the vital piece of the puzzle. Over the coming days other detailed analysis of imagery collected from Peenemünde finally revealed a small object sitting on a ski ramp. After all the heated debates over the differing interpretations of the incomplete intelligence picture, finally it was clear that the Nazis had developed a flying bomb. However, the intelligence picture that was emerging remained incomplete.

What was also becoming clear from ULTRA intercepts read at Bletchley Park was that the aiming accuracy of the V-1 was improving. Intercepts of the plots of the V-1 test firings at Peenemünde showed that the test firings in December 1943 resulted in a wide spread of impact areas off the Danish island of Bornholm. The scatter plots revealed by Dr R.V. Jones in his book

Most Secret War show a cross range and down variation in impact points of around 100 kilometres. By May the spread of impact points had reduced to a box with sides of 30 kilometres. These results were achieved in a series of over ten tests conducted between 6 and 10 May 1944.

The missile was assessed as having a maximum speed of 400 mph. This was very accurate. The analysis of the likely operating altitude lacked definition. A band of between 500 feet and 10,000 feet was too wide to help defence planners. After some further consideration it was suggested that the V-1 would probably fly at around 6,000 feet. This was double the actual operating figure. Early signs that this was an over-estimate also emerged from the analysis of the Peenemünde trials, where the operating altitude appeared to be lower.

In contrast, thanks to the accuracy of some of the intelligence received, the power plant and size and shape of the V-1 were well-known. All was now set for the battle that was about to occur. In a secret defence instruction issued at the time, a general classification was given to the V-1 missiles. They were to be known as 'Diver aircraft or pilotless planes'.

It was at this time that Flying Officer Barckley became the first pilot in the Royal Air Force to shoot down a V-1. It happened at night off the coast of France when a Tempest V of 3 Squadron came across a V-1 over the English Channel. Barckley recalled spotting a 'bright light' in the sky off Le Havre. His initial reaction was to think that a pilot had left his navigation lights on. He positioned himself behind it and opened fire. His log book records that he engaged a 'jet-ship' on the night of 8/9 May 1944. This was the first of a total of twelve V-1s that he was to claim. He also shared a claim on a thirteenth. This event was clearly not part of an orchestrated attack. It is likely it was a nighttime test firing aimed at validating the operation of one of the V-1 launch sites on the Cherbourg Peninsula. These were built to provide aim points in Bristol and Plymouth. Twenty-four hours later the second V-1 was brought down by a Beaufighter crew from 68 Squadron.

First sight

The analysis of the threat was discretely circulated to the members of the Royal Observer Corps who were going to be in the front line detecting any incoming missiles. Those who received the material were told of its sensitivity. The aim was to prepare them for what was clearly about to be unleashed on southern England. When the first one was spotted over the coast of Kent at 4.06 am on 13 June the two individuals (one a greengrocer and the other a builder) who saw the missile knew immediately that it was the Nazi secret weapon on which they had been briefed.

Barely a minute earlier two auxiliary coastguards also noted the arrival of two V-1 missiles from their location on the edge of the breakwater at Folkestone Harbour. They had not been briefed on the nature of the threat and recorded 'two aircraft with lighted cockpits coming in from the French coast in a north-westerly direction'.

The observation post of the Royal Observer Corps personnel was atop a Martello tower at Dymchurch on Romney Marsh. This gave them a good platform from which to initially detect the missile and track it as it passed overhead heading inland. The tower was one of 103 that had been built during the Revolutionary Wars 140 years earlier to counter an invasion from France. They were to be in the front line of the defence of Britain. Now, years later they were to be at the apex of a complex multi-layered defence system designed to reduce the numbers of V-1 missiles that would actually reach their main target, London.

As the missile passed overhead the two observers went into action. While one of them tracked its passage inland, the other quickly telephoned the message through to their headquarters. The report was simple but was quickly sent through an expectant chain of command. It read 'Mike-Two (their call sign), Diver, Diver, Diver one four, north-west one at one.' Twelve minutes later at Gravesend the V-1 crashed. The bombardment of London and southern England had started. The headlines in the *London Evening Standard* published on Friday 16 June read: 'Pilotless Planes now Raid Britain.'

Three months later in the *Kent Messenger* newspaper, the headline 'Kent had 2,400 Fly-Bombs: 100 more than London' made the reality of what had unfolded that June evening very clear. A map produced by the newspaper showed just where the flying bombs had landed in Kent. Very few areas were immune from attacks, with the area from Romney Marsh through Tonbridge and Maidstone up to Dartford being most heavily targeted. The success of the anti-aircraft guns based along the coast was also apparent in the map with markings indicating those V-1s that had been shot down into the sea. The area between Dymchurch and Folkestone was one where many V-1s ended their journey. A small cluster off the coast of Dover also showed how effective its anti-aircraft fire had been.

Reports of damage from V-1 attacks were noted in 448 communities as 1,444 missiles landed in the county. A total of 152 people died in Kent as a result of the V-1 attacks and 1,716 were injured. The village of Tenterden received 238 impacts. Ashford followed with 184 with New Romney at 149. This was the area that became known as 'Doodlebug Alley'. But the attacks were obviously not simply confined to Kent.

East Sussex was also an area badly affected by V-1 strikes. In total, records show 775 landed in areas such as Uckfield (146), Hailsham (159), Battle (374), Cuckfield (20), Bexhill (16) and Hastings (14). In total Sussex was to receive 880 attacks. Surrey (293) and Hampshire and the Isle of Wight (80) were to get off lightly.

As an indication of the bias towards the eastern side of the county of East Sussex no V-1 landings were recorded in Burgess Hill, Brighton, Hove and Portslade. The twenty V-1 strikes in Cuckfield were also very spread out geographically. Those that impacted did so away from built-up areas. The pattern of landings in East Sussex also reveals something of the impact of the interceptions by the Royal Air Force. Many landed in deserted rural areas. They had evaded the anti-aircraft batteries along the coast but succumbed to being shot down.

This area of East Sussex and Kent was where the main emphasis of the defences against the V-1 was operating. To the west of East Sussex, aside from interdiction by the Royal Air Force there was no other means of engaging V-1 heading for targets such as Portsmouth, Southampton or Bristol. On those occasions where the number of Royal Air Force fighters in the air was less than the number of inbound V-1s it is likely the fighters were tasked with engaging those missiles bound for London.

Along the coast on the Isle of Wight the patterns of attacks were slightly different. Over a period of nine nights from 5 July fifty V-1s overflew the island en route to attack Portsmouth or Southampton. With no anti-aircraft batteries to engage the missiles, they either flew on to their targets or crashed into the sea. For the ninety seconds the V-1 took to overfly the island, it was a difficult time. It is not hard to imagine people holding their collective breath. In the course of this short period two V-1 missiles fell on the island, one killing a person in the town of Lake. Once the Allies broke out from the Normandy beaches the attacks from fixed sites along the French coast moved. The Isle of Wight had endured a short but scary period and had come out of it relatively unscathed.

As the Allies continued along the French coast capturing the fixed launch sites the pattern of attacks shifted from East Sussex and Kent to Essex. The main thrust of the V-1 attacks from fixed sites was over by the end of August. While some sporadic attacks continued until March 1945, they were not of the same intensity. A V-1 that landed at 0535hrs in Dartford on 16 March did damage 450 houses. It showed just how large the damage radius could be in an urban situation. It was almost the last one to be recorded landing in Kent.

The strikes that then started to occur in Essex were mounted by the air-launched variant of the V-1, the land-based fixed sites now having been largely overrun by the Allies in France. From June 1944 to March 1945 Essex was subjected to 412 attacks, one of the worst being in Romford on 30 June 1944 when over 500 houses were damaged. For the people of Colchester the nights at the end of September 1944 became punctuated by the sound of incoming V-1s. By way of contrast, Suffolk (93) and Norfolk (13) clearly did not attract as much attention as Essex.

A second Blitz?

In the first week of the campaign 756 people would die and 2,697 were injured. This was indiscriminate warfare. Twelve schools and hospitals were also damaged, along with four churches. The worst initial incident was on Saturday 17 June when a V-1 struck St Mary Abbot's Hospital in Kensington. It destroyed the children's isolation block, killing thirteen children and five adults.

From the Nazi viewpoint the campaign got off to a stuttering start. Five of the ten V-1 missiles launched crashed shortly after take-off. The very first V-1 to land in Great Britain landed in Southampton. Only one of the first ten to be aimed at London reached what they had designated 'Target 42'. It landed at 4.25 am on a bridge over Grove Road in Bow. It carried four tracks of the London and North-Eastern Railway (LNER) from Liverpool Street. Nearby houses were also damaged in the explosion and eight people died with thirty seriously injured; another 200 were made homeless.

The other three landed in Gravesend, Bethnal Green and Sevenoaks. It was hardly an auspicious start. In fact only ten V-1 missiles were launched on the first day of operations. For some in the Cabinet who had reservations about the real capability of the so-called secret weapon this was an opportunity to ridicule the previous intelligence assessments.

That sense of euphoria, however, was short-lived. Two nights later a sustained attack started at 11.18 pm and lasted until noon of the following day. During that time 244 missiles were launched against London and a further fifty with Southampton as their target. Of the total launched, seventy-three reached the capital twenty-two minutes later.

Remarkably, given the situation, thirty-three had been shot down en route. Eleven made it through and reached Greater London. If the night had provided a point of calibration to those who sought to minimize the potential effects of the V-1, the events that occurred on the first Sunday of the campaign were to provide a stark portend of what was to come.

At approximately 11.00 am a V-1 landed on the Guards Chapel in Birdcage Walk. The explosion killed 121 people and injured another 141. Such was the scale of the attack that it was two days before the last of the bodies were removed from the site. On that same day the Nazis fired their 500 V-1s. The bombardment of London by the V-1 had just become very serious. In July 1944 just ahead of the peak of the V-1 bombardment 2,441 people were killed and 7,101 injured in V-1 attacks.

In the first two weeks of that month nearly 4,000 missiles had been launched against London. Out of 3,000 that reached the air defence systems around London, 1,192 were destroyed. Fighters accounted for 924 and 261 were blown up by the anti-aircraft guns. A total of fifty-five were also interdicted by balloons. The apparent initial success delighted the Nazi high command who immediately sought to expand the attacks.

Despite the initial onslaught, many questions remained in the minds of the defence analysts charged with trying to work out how the Nazis would use the V-1. What was the payload that the missile could carry? Over what range could it operate? How many could the Nazis build? How many might arrive per day over England?

With memories of the Blitz still fresh in people's minds, inevitable comparisons were made in the analysis. Over its 267-day window London particularly had been attacked on seventy-one occasions, with one period lasting for fifty-seven consecutive nights. This resulted in over 1 million homes being destroyed and over 40,000 civilians killed. Estimates of those injured during the Blitz varied from circa 50,000 to 140,000 civilians.

Major attacks on cities saw over 100 tons of explosives dropped by the Luftwaffe. In total throughout the Blitz 18,800 tons of high explosives were dropped on London. This gives an average of 125 tons (330 bombs) per raid. Could the V-1 campaign get close to this kind of level?

To assess this from a historical viewpoint it is necessary to know that the warhead on the V-1 weighed 1,900lb (862 kilos or 0.84 ton). It was made of amatol. This is a mixture of TNT and ammonium nitrate. The combination allowed the Nazis to maintain the overall destructive effect of the warhead while using scarce TNT efficiently.

In some scientific circles in the United Kingdom the idea that a flying bomb could carry a large warhead of this magnitude was dismissed almost out of hand. However, if the warhead was only quite small, what was the point of developing such a weapon? This sparked a range of rumours and speculation as to what the V-1 might actually carry.

In Washington at the time one rumour even went so far as to suggest that the V-1 could be modified to carry a biological weapon that could 'kill every living creature in the British Isles' using a fearsome weapon called the 'Red Death'. Another suggested that the missile could be equipped to carry radiological material which it would disperse over a wide area. The aim, that particular assessment noted, would be to make a 2-mile-square area uninhabitable.

While it is difficult to be precise about the total numbers of V-1s fired at England during the bombardment, several sources quote the number at 9,521. By contrast, however, the book written by the wartime head of British Scientific Intelligence, Dr R.V. Jones, called *Most Secret War* states that according to 'The Defence of the United Kingdom' that figure was slightly less at 8,617. Nevertheless, all sources agree on one fact. The initial stages of the V-1 bombardment from 13 June to 1 September saw 2,334 V-1 missiles land on targets across the capital. Others either fell short, were targeted elsewhere (places such as Portsmouth), or were successfully engaged by the air defence systems.

The average daily rate of explosives arriving over England in the early stages of the V-1 bombardment was therefore 25.2 tons or 20 per cent of the total delivered on average in the course of the Blitz. Yet even this was only sustained for a relatively brief period. This was hardly a rate of attack that merited being referred to by some commentators at the time as being the equivalent of a second Blitz. In fact, even referring to the V-1 attacks as a bombardment is arguably stretching a point. Official records show that 6,725 V-1s actually crossed the coast of England out of the total launched. This suggests a failure rate on take-off of around 22 per cent based on Dr Jones's assessment of the total number of launches.

That said, the V-1 could have been launched at a rate of fifteen per day from each of the ski sites. The first week of July saw the Nazis launch 800 V-1 missiles; the peak rate they achieved in the campaign. On 2 July 161 V-1 missiles crossed the English coast. Such was the mounting concern at this point that Prime Minister Winston Churchill made a statement to the House of Commons in which he stated: 'up until 6 am today 2,752 people have been killed by flying bombs and about 8,000 have been injured and detained in hospital.' He went on to note that in the same period 2,754 flying bombs had been launched against London. In his remark he was presumably not referring to the actual launch rate as he had no way of knowing that number. The figure of launches he quotes was in fact the number that actually landed in London. Again the figure of around one person dying per V-1 that landed comes across.

The maximum launch figure that was ever achieved from a single site in a day was eighteen. Film footage from the time shows operators manhandling missiles onto the launch skis. It was heavy work, involving a team of six people just to move the weapon and set it up on the ramp. This was quite a cumbersome process. That said, any analysis of the launch rates achieved does not show a level rate. While from the outset the rate of launches declined, there were periodic surges. Over the period from the start of the campaign to the beginning of September, a period of eleven weeks, ten of these notable increases in tempo exist.

Sustained rapid fire, even if the inventory of missiles was available, was simply not possible. Also the logistic supply chain was unable to move the numbers of missiles from the factories where they were built at a sufficient rate to maintain a sustained bombardment. Raids on the rail networks by Allied bombers were having some success in disrupting the movement of trains carrying the weapons to the launch sites. However, on the peak day of the bombardment on 3 August 316 V-1s were launched in a 24-hour period from thirty-eight launch sites. Twenty-five of these crashed immediately after take-off. Only just over 100 were actually recorded as having made it across the coast.

Had all of the approximately 100 sites been able to achieve this kind of launch rate, the situation could have been very serious. It is interesting to note that in one of the early intelligence assessments given to the Cabinet a figure of 1,500 missiles being fired each day was reported. While no definitive source exists as to the maximum number of missiles that were launched on a single day, the figure of around 100 is routinely mentioned in reporting from the time, although a single source suggests the total may have been as high as 190 per day. Other sources suggest the average rate of launches was between fifty-five and sixty per day.

The threat eases

As the Allies marched across Europe, so the Nazis had to move the launch sites and eventually resort to the use of the air-launched variant of the missile. This dramatically reduced the rate of their arrival. The last V-1 was launched from France on 1 September 1944. This ended what could be thought of as the first phase of the bombardment. From this point onwards, the Nazis either used the air-launched variant of the missile or ground-based sites in the Netherlands. Attempts to increase the range of the V-1 foundered.

If each V-1 that had been targeting London had bypassed the defences, the total tonnage of high explosives delivered would have been 8,000 tons, the equivalent of around sixty-four nights of routine bombing in the Blitz. This would have been quite a sustained level of attack. However, in-flight failures and a comprehensive multi-layered defence system swiftly assembled over southern England reduced those arriving over London to around one-third of the number launched.

As to estimates of the production capability of the V-1, these varied wildly. Early in 1944 an intelligence assessment published by the Joint Intelligence Committee estimated the Nazis could be capable of launching 45,000 V-1 on attacks against England. Had this been achieved and all of them got through the defences, this would have been the equivalent of twice the tonnage dropped in the Blitz. This assessment was openly challenged by those close to Winston Churchill. Among many who had doubts was Lord Cherwell, scientific adviser to Churchill. He specifically questioned where the Nazis could be manufacturing such vast numbers of missiles.

Historical records obtained after the Second World War suggested that the total number of V-1 missiles manufactured by the Nazis was around 30,000. If all of these had successfully been used to attack London it would have delivered 25,200 tons of high explosives onto the city. This would have been the equivalent of 134 per cent of the total dropped in the Blitz. However, the daily delivery rate would have been much lower. This would have stretched out the impact, which would have had its disadvantages as well as advantages.

In terms of the death toll, a comparison between the Blitz and the V-1 bombardment shows that the latter was not so effective as a killing machine. The V-1 attacks claimed 6,184 lives and 17,981 injured. These numbers are significantly lower than those of the Blitz. However, for the Nazis the lower monetary cost of the V-1 attacks was an advantage.

Interestingly, however, when analyzed on the basis of how many people died per ton of high explosives, the comparison between the Blitz and the V-1 bombardment is not dissimilar. These figures suggest that initial Nazi high command views that the V-1 would be a weapon that would turn the war were extremely optimistic. If the Blitz did not break the spirit of Londoners in terms of quantities of high explosives delivered, the V-1 was hardly going to be a game-changer. Even if they had been able to significantly enhance the production rates of the V-1, it is unlikely that some tipping-point would have been reached at which the outcome of the war could have been in doubt. It did, however, have a psychological effect. With the war drawing to a close the ability of the V-1 and V-2 ballistic missile to rain death upon London with little warning was nerve-racking.

Analysis of the impact of the V-1 bombardment

Across London, however, in the summer of 1944 there was a real sense of being in the cross-wires of the V-1. The boroughs of Croydon (141), Wandsworth (122) and Lewisham (114) recorded the highest number of attacks. As these lay on an arc from the major launch sites in northern France it suggests this was the main area targeted in London.

Pinpointing a specific aim point from the data is difficult, although some reports have suggested that the Nazis erroneously picked North Dulwich Station as their datum for the attacks. Dr R.V. Jones in *Most Secret War*, however, suggests that the central point of aim was Tower Bridge. His viewpoint is backed up by evidence obtained from the Nazi headquarters running the V-1 campaign when this was later overrun by the Allies.

This viewpoint is also borne out by an analysis of the main cluster of aim points. This occurs in a triangle bounded by the London boroughs of Westminster, Lambeth and Camberwell. Croydon was also one of the largest boroughs in London from a geographic viewpoint. So it might expect to have received the highest total number of raids. This spread on the ground is again borne out by overlaying the spread of the results of the V-1 test firings. Maps of the impacts of the V-1 in Croydon show that hardly a street in the borough escaped an attack. In his analysis of the pattern of attacks in the borough, author Bob Ogley notes that 58,968 houses were damaged. This is more than the total number recorded in the borough. Clearly some houses were hit more than once.

In terms of the worst day for a single area of London, a limited sample of the records from the time shows that on 22 July Beckenham received six V-1 attacks. This was twice the maximum figure for a single day of any other London borough south of the River Thames. Such was the level of concern that factory spotters were placed on the rooftops of major industrial facilities to forewarn people of an impending attack on their building.

One unnamed person kept meticulous records of the numbers of V-1 he heard as they passed over his factory in Esher. Situated 14 miles to the south-west of London, Esher was an ideal location to chart the sounds of V-1s heading along the main attack routes over Croydon into London which was 10 miles away, the distance over which the sound of the V-1 engine carried.

His records were published by Peter Cooksley in his book *Flying Bomb*. They show that on 22 June 1944 he heard forty-seven V-1 missiles. This was more than twice the figure heard on any other day in June. In July his records show the worst day as 21 July when fifty-six missiles were heard passing overhead. By August the rate of attacks he observed started to drop off. The worst day of that month was 6 August when eighteen missiles were noted in his log book. In total over the three-month period at the start of the V-1 bombardment, the observer recorded hearing 607 flying bombs in the seventy-seven days from 16 June to the end of August. This broke down into 169 in June, 290 in July and 148 in August.

What is also interesting is the variation in the daily totals. From 24 June to the end of the month, only twenty-nine V-1 missiles were heard. Another surge occurred at the beginning of July, when in the first four days his records note ninety-four V-1s. Another fallow period follows before the largest single day on 21 July. Looking at the records it seems this was the final major fling of the bombardment. In the remaining forty-one days of his records, the total only passed a count of ten on six occasions. Nazi records analyzed after the war showed that the busiest day was 3 August when 316 missiles were launched, 220 of which were to arrive over London.

The extended arc from Wandsworth through Lambeth (71), Camberwell (80) to Lewisham and beyond to Greenwich (73) and Woolwich (77) was where the majority of the V-1 attacks (537) occurred. This was 23 per cent of the total raids by V-1 flying bombs. This was also the location where the highest density of attacks occurred when the ratios are listed by the geographic area of the boroughs.

Maps produced of the sites where the V-1 landed show one cluster of fourteen V-1 craters within a distance of 4 miles of Lewisham railway station. A similar pattern occurred around Greenwich railway station with nine craters. The Royal Naval College received two direct hits on its grounds but the Museum remained unscathed. One V-1 landed close to the Royal Observatory. The majority (1,493) or 64 per cent also fell south of the River Thames. North of the river the pattern of attacks was more diffuse with the East End of London receiving a higher density of raids.

Out of the ninety-five boroughs in London at the time, these bore the main brunt of the attacks. Had the pattern of V-1 attacks been randomly distributed throughout London, each borough would have received twenty-five hits. Perhaps surprisingly, the City of London was only attacked on seventeen occasions. However, this is more likely to be a reflection of the overall accuracy of the V-1 flying bomb. Only twenty-nine of them landed in Westminster, although for its relatively small size this was quite a high overall density. The north-west of London was the safest part of the city.

In his book *The Doodlebugs: The Dramatic Story of the Flying Bombs of World War II* Norman Longmate estimated that over 5,000 families felt the pain of bereavement. More than 50,000 people were also disabled by V-1 attacks, with 723 losing their lives in the first week of the onslaught. By 15 July, one month into the campaign, he notes that the death toll was close to 3,600, rising to 5,476 by the end of August. From his analysis he documents 2,224 V-1 flying bombs that landed in London killed 5,126 people, an average of 2.3 per bomb. This differs slightly from other accounts of the period.

Longmate's analysis compares this with the casualty count emerging from the Blitz and notes that up to May 1941 a mathematically-derived figure of 0.83 people had died for each metric ton of explosives dropped by the Germans on London. They seriously injured 0.93 people per metric ton. By contrast over the full period of the bombardment the V-1, with similar explosive power, killed 1.1 people and seriously injured 3.1.

In southern England outside the capital, where many flying bombs fell short of their intended target, the death toll was approximately 350 or 0.1 per bomb. The figures for serious injury were 6.6 per bomb in London and 0.4 per bomb outside. Longmate quotes Prime Minister Winston Churchill making the point that each flying bomb killed on average, across the campaign, one person. It was a statistic that was to hold true until the onset of the more devastating V-2 attacks.

Analysis of the arrival times of the V-1 flying bombs in the first phase of the campaign from June to September 1944 shows that the Nazis slightly favoured nighttime launches. Based on a sample of 465 records collected from several sources over the period from 13 June to 1 September, V-1 attacks are recorded around London throughout the day. However, there is a small but detectable bias towards attacks occurring between midnight and six o'clock in the morning. The low points occur around midday and in the evening.

In Wandsworth, Lambeth and Croydon the time between midnight and two o'clock in the morning saw a peak in arriving missiles. On a number of occasions the attacks occurred almost simultaneously as a result of salvos fired from several launch sites in northern France. The implications of this are that to a first order the bombardment was maintained throughout the day with launches occurring from the ski sites in France at around two per day.

The area where the pattern of V-1 attacks was at its greatest included Kent (1,444), Sussex (880) and Essex (412). A secondary ring of attacks at a lower level included Hampshire (80), Hertfordshire (82) and Suffolk (93). A tertiary ring of counties from Berkshire (12), Oxfordshire (4), Buckinghamshire (27),

Bedfordshire (10), Cambridge (8) and Norfolk (13) also saw attacks. There are no records of V-1 attacks ever occurring in Dorset, Wiltshire or Gloucestershire. This laydown of attacks across the south-east of England suggests that the combined effects of the prevailing winds and navigational errors may have had an impact on the V-1s, driving them off course.

The air-launched campaign

It was not only London and the south-east that were affected. The V-1 could quite literally fall out of the sky anywhere. Records show over the period from 13 June 1944 to 29 March 1945 that at least seven reached Yorkshire and eight landed in Lancashire. Given the limitations on range of the V-1 flying bombs based in France (a maximum distance of 163 miles with a fuel capacity of 640 litres), the weapons that landed in the north of England were likely to have been launched from bombers flying at low altitudes over the North Sea.

The air-launched element of the V-1 campaign was due to get under way as part of what Hitler called Operation *Eisbär* (POLAR BEAR) in June. As a result of delays the air component did not start its attacks until 9 July. Aircraft involved took off from bases in The Netherlands. Over the coming weeks it became apparent that the air-launched component was nowhere near as accurate as the ground-based missiles. This, perhaps, is understandable.

The launch point for the ground-based missiles was accurate, as was the heading on which they were initially fired. By contemporary standards, aerial navigation was still in the Stone Age. Missiles aimed at Portsmouth landed in Southampton. Looking at the data, the War Office concluded that the air-launched version was around three times less accurate than the ground-based missiles. They also flew at slower speeds which increased their vulnerability to interception.

The air-launched element of the V-1 was almost an entirely nocturnal affair. Concerns over the vulnerability of the He-111 H-22 proved correct, even though they were equipped with the Liechtenstein radar-warning receivers. The Mosquito proved particularly adept in the role of hunting down the launch aircraft.

Records from the time show that 865 of the air-launched V-1 missiles were used between 16 September 1944 and 14 January 1945, a period of 120 days. These were fired by specially-modified He-111 H-22 bombers flown by the Kampfgeschwader 53 (KG53) 'Legion Condor' bombing wing. They had been moved from operations on the Eastern Front. Throughout the campaign the average launch rate of these air-launched weapons was ten per day. In October KG53 was able to muster seventy-seven bombers. A further twenty-four were in maintenance and repair.

The rate of launches was lower in December 1944 with no interceptions being recorded in the first two weeks of the month. This was because some of the air-launched missiles had developed the habit of detonating prematurely. A dozen bombers were lost at the beginning of December to this kind of incident. It created a pause in the campaign while the problems were sorted out.

This was a lull before a brief storm. On Christmas Eve KG53 launched its largest-ever formation of fifty bombers. Their target was Manchester. The raid was known as Operation MARTHA. Of the fifty missiles launched, only thirty reached the English coast. Fifteen of the missiles flew on towards Manchester but only a single V-1 landed in the city. It was a reminder of just how flexible air power can be in reaching areas that would otherwise have been well beyond the range of the ground-based missiles.

By the end of the campaign the nightmare of the V-1 attack had reached every region in England. One of the last major raids on the country occurred on the night of 3 January 1945 when forty-five V-1s were launched from their He-111 H-22 bombers. Twenty days later the aerial element of the campaign came to an end. Nazi Germany was rapidly running out of fuel. The air-launched V-1 campaign had lasted 178 days.

The final air-launched bombing attack on England took place on 14 January when twenty-five V-1 missiles were launched. Only seven reached their target. It has also been reported that 1,776 such launches took place in total over the extended period of the bombardment. Radar systems tracked 1,012 heading for England. From that total 404 were shot down and 388 impacted in England, 66 of those on London.

The total air campaign comprised around 10 per cent of the total V-1 missiles fired. At the end of the campaign the KG53 Legion Condor had lost seventy-seven of the modified Heinkel bombers. At least sixteen of those were lost to Mosquitos. As Steven Zaloga notes in his book *V-1 Flying Bomb 1942– 1952: Hitler's Famous Flying Bomb* these figures meant that only 4 per cent of air-launched missiles reached their target, and for each missile that did explode, a bomber was lost along with its air-crew. It was an extremely poor military return given the levels of investment made in blood and treasure.

Spies, ruses and deception
The whole question of trying to deceive the Nazis over the effectiveness of the V-1 bombardment was vexing. Double-agents operated by what was

known as the Twenty Committee (double cross or XX in roman numerals) were one of a number of means used to try to fool the Nazis into believing the V-1 missiles were overflying the centre of London. The idea was to try to deceive the Nazis into shortening the range of the missiles to compensate. One of these agents, code-named Tate, was used to pass on false information about where the V-1s were landing. Other agents who were being controlled by MI5 were also contacted by the Nazis and generated similar reports.

While the Twenty Committee fed back its false reports through its agent network, one obvious concern was that high-flying Luftwaffe photo-reconnaissance planes would somehow show the scale of the deception and also compromise the double-agent network. Reporting after the war showed that the Luftwaffe had been unable to conduct any flights over the London area from 10 January 1941 until September 1944. This was a three and a half year period when photographic intelligence was completely cut off from the Nazi war machine.

Flights only resumed when the Messerschmitt 262 fighter jet entered service. Its speed gave it an advantage over the traditional fighters employed by the Royal Air Force until the introduction of the Gloster Meteor. This was a remarkable situation. Fighter Command had been highly effective in intercepting Luftwaffe reconnaissance sorties. For them, the duration of the V-1 bombardment was a period when they had a limited ability to cross-check the reporting from the double-agents.

In many ways this was fortunate. Paradoxically when the Messerschmitt 262 did make some flights over London, cloud coverage over the area south of the River Thames prevented the Nazi photo-interpreters making any detailed assessment. North of the River Thames the analysis also mistook craters left from the Blitz as being V-1 impacts. This led them to believe, somewhat fortuitously, that agent reports of the V-1 flying too far were correct. This contradicted the assessments in the Nazi high command responsible for the V-1 campaign. Unknown to British intelligence, around 3 per cent of the V-1s carried a small radio beacon that allowed their trajectory across southern England to be tracked up until the point of impact. When the chips were down, however, the high command in Berlin believed the reporting from the agents.

In truth many of the double-agents involved in the main D-Day deception operation called Operation FORTITUDE had already given false reporting that should have compromised their cover. Intriguingly, the Nazis continued to place a high degree of faith in their double-agent system, almost refusing to believe it had been compromised. Two agents were held in particularly high esteem in the Abwehr. One was code-named Garbo and the other Brutus. Such was the faith in both men and their imaginary network of spies that the Abwehr handlers asked both men to collate reports on the V-1 impact areas.

The dilemma for MI5 was that if both men reported the results inaccurately, they may be exposed as double-agents. For the period from 16 to 24 July Garbo went off the air. It was uncharacteristic behaviour and had to be explained to maintain his cover story. Ingeniously Garbo invented a plausible story about being concerned about the accuracy of the V-1 reports that he had received from his network of agents. He reported that the reason he had been off the air was that he had personally decided to go and visit the sites to corroborate the information that he had received. In what was a complex piece of deception, another agent then reported that Garbo had been captured by the police.

This helped provide a plausible story concerning the lack of reporting generated by Garbo. When he re-established contact two weeks later, he reported his arrest but was able to reassure his handlers that he had managed to convince the police of his bona fides. Worried that any further visits might expose Garbo to yet further scrutiny, his Abwehr handlers decided it was too risky and asked him to stop reporting on the V-1 landing sites. They also asked Brutus to stop any similar reporting.

The ruse, however, was something that was hotly debated in Cabinet. There were two schools of thought. One took a pragmatic view that it was important to try to seduce as many missiles away from the capital as possible. The other side suggested that this was *in extremis*, playing God; somehow deciding that the people of rural Kent were less important than those in Croydon.

The deception activity was part of a wider effort aimed at countering the V-1 called Operation DIVER. The name of the operation was derived from the British code name for the V-1. As soon as the intelligence picture on the threat posed by the V-1 became clear, a number of measures were put in place to reduce its effectiveness.

Defences against the V-1

The challenge was to find a way of defending against a fast-moving target whose total flight time from take-off to detonation to cover a range of 130 miles – the figure initially used by the Nazis – was around twenty to thirty minutes. This is a figure slightly higher than a simple mathematical analysis of speed and distance would suggest. It arose because the V-1 missile initial acceleration up to its operational speed was sluggish.

The flight trajectory of the V-1 at altitudes of between 2,000 and 3,000 feet placed it in a corridor of reduced vulnerability to anti-aircraft fire. At lower altitudes the V-1 would have been vulnerable to light guns such as the 40mm Bofors. Had they flown higher they would have come into the optimum range of the heavier guns. Flying at around 400 miles per hour, the V-1 would traverse the engagement zone of the anti-aircraft batteries too quickly for the sighting of the guns to be effective.

The defence against the V-1 attacks was conducted under the name Operation CROSSBOW. This covered all aspects of trying to defeat the V-1 from its point of manufacture to its impact over a target area. Bombing the factories that made the V-1 depended upon accurate intelligence. It also risked the workers who had been forced into building the weapons, many of whom were already suffering the privations of concentration camps.

Attacking the launch sites may appear easy on paper but in fact their shape and length made them really difficult to target specifically. Carpet-bombing of the area in which they were located therefore had a limited impact. If the V-1s were to be seriously attacked it had to come from the point at which they were launched until the point of detonation.

The solution to this problem was to build a layered defence system in southern England. The first layer comprised fighters operating out of bases that were as close to the launch sites as possible. While the fighters faced little in the way of opposition from the Luftwaffe, air defences around the launch sites were extensive. The obvious place to intercept the V-1 was after it had climbed away from its ski ramp and settled on a steady altitude and heading.

To do this in the early phases of the V-1 bombardment the Royal Air Force and some units from the United States Air Force mounted standing patrols over the coastline of Kent, Sussex and Hampshire. In order to catch the V-1 these patrols had to be conducted at a higher altitude to allow the fighters to dive and gain speed on their prey. Initially only thirty Hawker Tempest aircraft were deployed on Operation CROSSBOW. Their need to provide a continuous airborne presence stretched their resources to the limit.

Mounting standing patrols over the coast over the twenty-four hours of the day was hard with such meagre resources. The patrols were flown at an average height of around 5,000 to 6,000 feet. This provided sufficient height to ensure the aircraft could dive and catch up with the target while not being too high to avoid seeing the flames from the engine.

On the basis that each standing patrol lasted for two hours and that six aircraft were airborne in three groups of two to cover the main threat corridors, each air-crew would be expected to fly a minimum of three two-hour sorties per day without any rest periods. Quickly this was increased to eleven squadrons, two of which operated the Mosquito night-fighters. Over time that was further increased to twenty-three squadrons. Fifteen of these operated single-seat fighters and eight the Mosquitos; the latter taking nighttime duties.

This provided a very basic first layer of defence. It was hugely vulnerable to tactics involving salvo launches where a number of V-1s were fired from different ski ramps in concert, the aim being to ensure several of them arrived over the target area within a narrow window of time.

While this was a valid tactic, analysis of the launch rates actually achieved by the Nazis suggests that rarely were they able to overload the layered defence. The peak arrival rate of V-1s meant that on average around five were arriving per hour over southern England. For the first layer of the defence the problem was to ensure they had enough aircraft airborne to provide the geographic coverage of the corridors along which they flew.

At night detecting the V-1 also proved troublesome, despite the flare from the jet engine. Fighters being vectored towards a V-1 that had been detected by the Royal Observer Corps were aided by the simple measure of firing 'snowflake' illuminating flares as a marker to where the report had originated. As time was of the essence, once a V-1 had been spotted it was a valuable way of helping the fighter pilots. This activity was known as Operation TROTTER.

It is also important to remember that at the time the main effort of the military was focused on France. Anything that was drawn away in defence of the United Kingdom hampered the advance through Europe towards Germany. However, as the impact of those V-1s that did reach their targets was felt, it became a priority to step up the resources allocated to the fighter defences. By the beginning of September this had grown to over 100 aircraft. This also allowed a second layer of air-to-air intercepts to be created close to London.

Cues passed from the Royal Observer Corps were routed by radio to the fighters. They usually operated in pairs. The standing patrols were mounted throughout the day. At any one time it seems likely that six to eight aircraft were airborne. When vectored onto the target, if cloud conditions permitted they would see the long plume emitting from the pulse-jet engine. At the start of the bombardment the intercepts by the Allied air forces were not particularly successful. Tactics had to be adapted and over time these improved.

Chasing a V-1 was one thing. Destroying it was quite a different matter. At the start it took an average of 500 rounds of ammunition to bring down a V-1. After eight weeks this had reduced to around 150. On occasions the outcome was not as intended. In Kent the very worst death toll of the entire V-1 campaign occurred when one that had been shot down crashed into Newlands Military Camp at Charing on 24 June, killing forty-seven people and seriously injuring twenty-eight. The impact had occurred a few minutes before breakfast was to be served.

One novel technique employed by Royal Air Force pilots was to manoeuvre alongside the V-1 and try to tip it up on its side. The aim of this approach was to make the V-1 dive into the ground, ideally in an open area. No actual contact was made between the two aircraft as the tipping force was created by the airflow going over the wing of the interceptor. Despite the stories surrounding this manoeuvre, it was not an approach used that often. Only three V-1s are claimed to have been destroyed in this way. The usual approach for the Tempest was to get around 275 metres behind the V-1 and fire its 20mm cannon. It was, however, a hazardous position to be in when the V-1 exploded. It is interesting to note that none of the top fighter aces in the Royal Air Force from the Battle of Britain ever claimed a V-1 kill.

The second layer of the defence system was the anti-aircraft guns placed on the coast. Originally these were to be located in a defensive ring around London on the North Downs. Concerns about the effectiveness of Nazi radar-jamming had driven this decision. However, as the war had gone on, this problem had been largely eliminated.

A revised plan was therefore created that saw an initial move of 800 guns, 60,000 tons of ammunition and stores and 23,000 men and women moved to a forward area near the coast over a period of forty-eight hours. In total the move had seen the vehicles of the Anti-Aircraft Command travel a total distance close to 3 million miles. It was an amazing logistical feat given the circumstances.

The plan saw 1,332 anti-aircraft guns deployed to protect the Solent, London and Bristol. Hitting a V-1 using an anti-aircraft gun proved not to be straightforward. On average at the start of the bombardment it took 2,500 shells to destroy a single missile; a figure that compares well to the 30,000 anti-aircraft shells it is generally accepted were fired to destroy a single Luftwaffe aircraft during the Blitz. By the end of the campaign that had reduced to one V-1 destroyed for every 100 shells fired.

There were two reasons for the improved success rate. The first involved the introduction of a gun-laying radar system with sufficient accuracy to track the V-1, estimate its flight path and then provide a cue to the anti-aircraft guns. The radar system had been developed in the United States by scientists at the Massachusetts Institute of Technology and was referred to as the SCR-584 automatic gun-laying system.

Radar development in the United States benefited from a close collaboration with the United Kingdom. The Tizard Mission (named after Henry Tizard, chairman of the Aeronautical Research Committee) travelled to the United States in September 1940 to share a number of new technological advances being pioneered in the United Kingdom with the Americans. One fruitful area where exchanges were immediately valuable was in the recent development of resonant-cavity magnetrons.

The United Kingdom had built one to work at the wavelength of 10 centimetres (3 GHz) and its power output was far superior to the existing American designs. Combining this with a similar American naval development created a solution that was to have a profound impact on Operation CROSSBOW. The SCR-584 operated at a pulse width of 0.8 microseconds and a pulse repletion frequency of 1,707 pulses per second. Its peak power output was 250 kilowatts and its frequency moved across four bands close to 3 GHz.

It had two modes of operating. The search mode was conducted in a conical scan. In this mode the radar looked for a high signal return associated with a target. Once the target was detected, the antenna would move into a tracking mode where servo-systems would automatically ensure the radar stayed locked onto the threat. What emerged from the British/US collaboration was a radar system capable of detecting large targets at a range of close to 40 miles and initiating an auto-track when that closed to below 18 miles.

Given the comparative sizes of the V-1 and a Heinkel-111 to a first order the radar cross-section of the missile would have been around half that of the bomber. As the detection range scales according to the fourth root of the reduction in the radar cross-section it would have had a minimal impact on the performance of the SCR-584 against a V-1. If the V-1 flew directly overhead the radar it would have been tracked for a maximum of around three minutes, during which time the anti-aircraft guns were brought to bear.

This was helped by the development of the first gun-laying analogue computer. Developed by Bell Laboratories, it was called the M9 Director. It was able to control four guns. The first demonstration of the system took place on 1 April 1942. The test was so successful that Bell Laboratories received a contract for the development of 1,200 devices on the following day.

In operation the M9 Director was often placed in a central position with the guns at four corners of a square. In the first (pre-radar) variants, operators would use telescopes to manually track the target and as they moved their sights the rotation of the hand-wheels was conveyed by an electrical signal to the director which then calculated a firing solution for the guns. The major advance with the M9 Director was that this was one of the first to come into service with a live feed from a radar system and the analogue computer system. Its introduction into service alongside the anti-aircraft batteries tasked with engaging the V-1 had a profound impact on the effectiveness of the overall defence system. Over 1,000 of the V-1s were shot down in the English Channel before they crossed the coast.

However, despite the simple track taken by the V-1 and the ease with which its next position could theoretically be estimated (its speed was to all intents constant), technology available at the time could not ensure that the anti-aircraft guns could actually hit the target and destroy it. What was needed was a way of getting close enough to the V-1 and then using a fragmenting warhead to explode close enough for large fragments to hit the missile.

The second important development therefore involved proximity fuses. As the anti-aircraft shells got close to their targets they detonated, showering the V-1 with fragments. In the first week of the bombardment the anti-aircraft defences accounted for 17 per cent of the V-1s that crossed the English Channel. By the start of August the guns were shooting down 60 per cent of those that penetrated the initial fighter screen. By the end of the month this had grown to 74 per cent. The introduction of these two technologies proved decisive. Towards the end of the bombardment the anti-aircraft defences would destroy a V-1 for every seventy-seven shells fired. This was a considerable improvement on their original level of success.

The third layer of the defence against the V-1 was another air defence layer mounted by aircraft. The final layer was 1,750 tethered balloons, although the edge of the V-1 wing was designed in such a way as to cut any cable it encountered. Despite this, balloons are reported to have accounted for 300 V-1 missiles.

Twelve squadrons from the Royal Air Force and one of the United States Fighter Groups (363) were tasked with intercepting and destroying the V-1 missiles as they flew towards their targets. During peak days the Nazis launched over 100 missiles, so the squadrons were kept busy.

By a stroke of good fortune the V-1 campaign had been launched at the height of summer. This gave the pilots longer daylight hours. When dusk did intervene, the plume from the V-1 engine was easy to see with the naked eye.

The engine could also be heard from nearly 10 miles away but at the speed of the V-1 that distance was covered in less than two minutes.

Variable weather did occur in the summer and code words were sent out each day that dictated how the air defence system was to operate. If the codeword was 'Flabby' the fighters were given priority over the anti-aircraft guns to avoid the risk of collateral engagements. When the codeword was 'Spouse' it was the fighters that had to keep clear of the coastal area where the main concentration of anti-aircraft batteries was located. This was usually sent when weather conditions would have made it hard for fighters to engage the missiles. The anti-aircraft guns were still able to spot the V-1 on their radar plots. Medium weather conditions were announced by the transmission of the codeword 'Fickle'.

Little has been published on the combat posture of the Royal Air Force during this time, other than the squadrons tasked with intercepting the missiles. It is possible to speculate that the radar cross-section of the V-1 made it more difficult to detect using radar. It was, after all, much smaller than the classic raids that occurred during the Battle of Britain.

What seems certain is that the Royal Observer Corps played a hugely important role in detecting the threat and passing the information through the chain of command to cue interceptions. The distinctive audio signal and the glow created by the plume from the jet engine would have enabled the V-1 to be detected. With the V-1 flying a constant heading, establishing a track for an intercept was not difficult. Time, however, was a precious commodity as it had been in the Battle of Britain.

So it seems likely that the Royal Air Force had to mount standing patrols to try to engage the V-1. At night the patrols were mounted by Mosquito aircraft from 605, 219, 264, 96, 419 (RCAF) and 456 (RAAF) squadrons. By day the task was taken up by Spitfires and Hawker Tempests. Towards the end of the campaign, records do show that some Hawker Tempest squadrons also flew at night, although their success rate was lower than the Mosquitos. This may be simply down to the radar on board the Mosquito and the fact that it was a proven night-fighter.

One of the main bases at RAF Newchurch was located as far forward as possible to ensure it had time to engage and intercept the missiles after they were detected. The danger of collateral damage was all too real. This would have created a narrow band in which the intercepts could have been achieved, given the relatively small differences in speed between the interceptors and the V-1.

Two of the squadrons (Number 91 and 322) were equipped with the Mark XIV Spitfire. This was powered by a Rolls-Royce Griffon 65, a supercharged

V12 engine that could generate 2,500hp at 8,000 feet. Its maximum speed was 448 mph; 70 mph faster than the earlier Mk V Spitfire. This enabled the Mk XIV to catch a V-1 that flew at 400 mph. They were to achieve a ratio of kills per sorties flown of 0.1 or 9.5 sorties per kill.

In addition to the Spitfires the Royal Air Force also deployed seven squadrons of Hawker Tempest V aircraft. The top speed of the Tempest was just below that of the Mk XIV Spitfire at 442 mph. While its rate of climb was better at 4,700 ft/min, for the low-altitude engagements against the V-1 this did not have a marked impact on its ability to engage the missile although height was often an advantage at the start of an engagement as the intercepting aircraft could build up speed to catch the V-1.

The excellent low-altitude performance allowed the Tempest V aircraft of 3 Squadron, Royal Air Force to claim 288 V-1s shot down while they were based at RAF Newchurch in Kent. This was a temporary airfield created on the south coast for Operation DIVER. In total the 150 Tempest Wing of around 100 aircraft claimed 638 of the total number (1,771) of V-1 missiles shot down by aircraft. They were to achieve the highest ratio of kills per sortie at 0.21.

One Tempest pilot, Squadron Leader Joseph Berry from 501 Squadron, shot down a remarkable fifty-nine V-1 missiles. On one night alone he accounted for seven. In an effort to reassure the public, he was asked to go on BBC radio and give an account of a typical day in the life of a fighter pilot trying to shoot down the V-1. His sister, commenting after the war, noted the tempo at which the pilots had to fly. She noted that he visibly aged during the campaign: 'There were times when the aircrew were flying virtually non-stop for hours on end.' In one incident Squadron Leader Berry fell asleep while flying, only to wake up just before his aircraft collided with a balloon. The commander of 150 Wing Royal Air Force (Wing Commander Roland Beamont) also made a significant contribution, adding a further thirty-one kills to the total. He had shot down his first V-1 on the night of 22 June.

The Mosquitos of 96 Squadron, Royal Air Force claimed a further 428 V-1s with the Spitfires noting 303 kills. The Mosquitos achieved the first successful engagements of V-1 missiles on the second night of the bombardment when four were shot down. Three fell to 605 Squadron and one to 219 Squadron. During July and August Mosquitos from 418 Squadron took part in a number of patrols looking for V-1 missiles.

Wing Commander Russ Bannock led from the front, being credited with nineteen successful V-1 kills. His co-pilot recalls flying eighteen missions looking for the V-1 between July and August 1944. They were interspersed with missions designed to interdict other military targets in France. The Mosquitos also had some success against the air-launched variant of the V-1 by shooting down the He-111 H-22 on which the missile was carried to its launch point. Twenty Mosquitos would be lost during the V-1 bombardment. The pilots and navigators from 68 Squadron paid a particularly heavy price, losing six aircraft and twelve airmen. This was one-sixth of the total number of air-crew lost in the V-1 attacks.

In the book *Mosquito Missions* by Martin Bowman, Wing Commander Russ Bannock recalled the difficulty of the operation as the Heinkel flew low over the sea to avoid being detected by the radar systems on the Mosquito. The Heinkel mission often started out from the coast of The Netherlands from a radio beacon at Den Helder. Flights could last between three and five hours. At the release point the Heinkel pilot would quickly climb to around 1,600 feet, fire the missile and then quickly return to the comparative safety of low level. The Mosquitos, however, adapted their tactics and would lay in wait as the Heinkel tried to land.

In total 4,621 V-1s were destroyed by anti-aircraft guns and fighter intercepts. Two Mustang III squadrons (315 and 129) were also committed to Operation DIVER. The Mustang III was able to achieve a top speed of 417 mph at 2,000 feet which made it a useful platform to chase the V-1. They claimed 232 of the total V-1 missiles destroyed.

The arrival of the Gloster Meteor into service at the end of May 1944 was timely. It was the United Kingdom's first operational jet aircraft. The first squadron formed was 616 which arrived at RAF Manston in east Kent and rapidly worked up their operational capability. On 27 July a flight of four aircraft was declared operational.

At 14:30 on that day the first Meteor I took off on an operational flight to patrol above Ashford in Kent. While no V-1s were seen, over the coming days a number of encounters occurred that suggested the Meteor would be able to make a contribution to the overall defence against V-1 attacks. It was to be a brief but slightly frustrating period for the small group of pilots of 616 Squadron. It seemed that everyone else was having fun shooting down V-1 missiles by the score. However, it would not be long before 616 opened its score.

The squadron claimed their first kill on 4 August when Flying Officer Dixie Dean managed to tip a V-1 on its side when his guns had jammed. Within an hour 616 had claimed its first orthodox kill of a V-1 when Flying Officer Jock Rodger engaged a V-1 with guns. In total, records show that the

Meteors achieved 12.5 kills during 260 sorties. The reference to half a kill is due to one being shared with a Hawker Tempest.

This sounds like a low figure but one of the reasons was that the aircraft was so new into service that it could only fly for just under an hour on each mission before needing to be inspected. Meteors therefore did not take part in standing patrols. This reduced the number of possible targets they could have destroyed. At this time the rate of V-1 launches was also dropping off as the Allies moved across northern France.

The pilots of 616 Squadron also had some close shaves with their own side. It seemed that the defences saw anything that was jet-powered as a threat. On one occasion a Meteor was bounced by two Spitfires. On another, one was engaged by anti-aircraft batteries. The problem of what today is known as collateral damage did have an impact throughout the V-1 bombardment. Eight crew members were to lose their lives as a result of collateral fire. Two Mosquitos were brought down but one navigator managed to escape from the aircraft.

What is clear from all this analysis is that the cost to the defence outweighed by some margin the investment the Nazis had made into the V-1 programme. A report published by the Air Ministry on 4 November 1944 suggested that the ratio of costs was four to one. For every unit of currency invested by the Nazis, the United Kingdom had to invest four. Looking back on the analysis it is likely to be an underestimate of the reality of the relative expenditure involved as the Air Ministry confined its cost base to the investments in defence rather than looking at a wider measure of the economic impact of the destruction caused by those V-1s that penetrated the defence system. That said, the defences against the V-1 did prove important. As in the Battle of Britain, it helped to lift civilian morale. However, and this is not to sound churlish, it is important to remember that the V-1 was a relatively easy target to attack. By contrast, the V-2 ballistic missiles simply could not be engaged. The solutions to attacks by ballistic missile are only now starting to be addressed some seventy years later.

CHAPTER 4

Target Practice

Looking back at the applications of UMA in the Second World War it is easy to see why this was a time at which they became an established part of the military inventory. That was the legacy of the war. Yet during the war the UMA were confined to quite a narrow role as a first generation of guided missile whose accuracy was not that important.

For the Nazis the V-1 weapon was all about revenge for the day-to-day bombing strikes on Germany. As they could no longer mount manned retaliation, they would resort to developing an unmanned response. That carried extra weight due to the way in which the V-1 missile system operated. It added a new psychological aspect to warfare. In total war, when some of the constraints of contemporary warfare were simply not a consideration, the V-1 and V-2 truly became weapons of terror.

The lasting legacy of the Second World War as far as UMA were concerned was that they could be launched, maintain a steady heading and altitude towards a target and decide the point at which they would make their attack. Communications with the first generation of UMA was very limited. For the V-1, a very small number tailed radio antennas to help the Nazis plot their flight paths, albeit somewhat inaccurately.

The next and most obvious step in the development of UMA was always going to be a situation where they could be controlled remotely via a radio system. The issue for that, of course, was that once the UMA was over the horizon the ability to control it quickly disappeared. Today's UMA rely heavily on satellite communication links to bypass the limitations of a radio horizon. In the 1950s that simple limitation also had an impact on the various roles that UMA could perform. One obvious one was as a target drone.

Target drone developments in the United Kingdom
Initial work on the development of a target drone in the United Kingdom had started at the RAE at Farnborough in England in October 1930. To test the

ideas a Fairey IIIF aircraft was specially modified to augment the existing flying controls to facilitate remote operation. It was known as the Fairey Queen. Initial trials with a crew aboard were followed by a move to Lee-on-Solent where the aircraft was fitted with floats. Three test aircraft took part in a series of trials to evaluate its role as a target drone.

The test aircraft was then embarked upon HMS *Valiant* on ski launch for gunnery trials in 1933. After analyzing the results of the trials it was decided to embark upon an exercise of developing a dedicated target drone aircraft. It had, however, been the start of a series of developments of target drones, all of which would have the name 'Queen' associated with them.

What emerged from that exercise was a hybrid. The final target drone platform was based upon a modified Tiger Moth airframe that included elements of the original D.H.60 fuselage from which the final design of the iconic Tiger Moth was based. It was called the Queen Bee and it first flew in England in 1935. This became the world's first target drone aircraft. A total of 420 were built, of which around 380 were employed by the Royal Air Force as targets for anti-aircraft defences ahead of and during the Second World War.

Going back to the roots of the development of the Fairey Queen, the Queen Bee was also to serve an important role as a target drone for the Royal Navy. It was about to learn some harsh lessons about air power during the Second World War that would stimulate the development of the next generation of target drones.

At the end of the war developments of target drones in the United Kingdom would initially focus on a joint project with the Australians. The aim was to build a target drone that could be used during guided missile-testing. The Australian Government Aircraft Factory (GAF) had already developed two prototype target drones. These were built to test the aerodynamics, engine and radio control systems. They were called Pika which is the Aboriginal word for 'flier'. Its first flight took place in 1950. Out of this proving model a new target drone was to be jointly developed. It too was named using an Aboriginal term; in this case aptly called Jindivik, 'the hunted one'.

Over the lifetime of the platform more than 500 were built. At the United Kingdom test facilities at Llanbedr and Aberporth Jindiviks became a routine sight up until the turn of the century. They flew alongside target tugs such as the Westland Lysander, the De Havilland Mosquito, the Gloster Meteor and the English Electric Canberra. The Canberra was eventually replaced in the target towing role by a BAE Systems Hawk. When Jindiviks flew on the test range they were often escorted by a manned fighter jet in an observation role.

The Gloster Meteor towing tug was designated the TT20. Twenty-six aircraft were modified to house a wing-mounted winch that could pay out over 6,000 feet of cable at the end of which was the target. On take-off the target was housed in the rear of the aircraft. Three other types of aircraft were also flown as target drones. These were two former Fleet Air Arm Fairey Firefly aircraft that had been retired in 1956, three Gloster Meteors (U14, U15, U16) and two De Havilland Sea Vixens.

The conversion of Gloster Meteor to the role of target drone fits a pattern adopted by many other countries. As jet aircraft are retired from service a number of the airframes are converted to become target drones. Their manoeuvrability adds realism to the types of engagements that can be used to evaluate new missiles. Initially ninety-four of the F4 Gloster Meteors were converted to become target drones. Extra generators had to be added into the airframe to provide the power for the payload and test equipment needed to operate the target drone from the ground. Their designation was U15.

Fifty-nine of these were sent to the Woomera Weapons Research Establishment in Australia. Twenty-three were based at Llanbedr. The last Meteors used in the target drone role were designated the U16 and around 150 of these were built. This mix of former manned fighter jets working alongside UMA was a model many countries have followed in their development of target drones.

Conducting realistic testing of missile systems has required target drones to evolve in two areas. The first has been the need to ensure that target drones can fly manoeuvres that are indicative of the latest generation of fighter jets. As they have evolved to exploit the latest research in aerodynamics and flight control systems, so the target drones have also had to develop. The second element concerns the defensive aid suites on contemporary fighter jets.

As missiles have moved away from command guidance from a launch platform towards increasing autonomy of sensor systems in their seeker heads, the defence of the fighter has had to involve both manoeuvring and deploying countermeasures. This is a far cry from the days when target drones had to enhance their signatures in ways to help them be successfully targeted. While carrying that form of payload is still important, such as when testing a missile's ability to follow a manoeuvring target and get sufficiently close to have a high probability of a kill, other payloads also need to be carried.

Chaff, infrared flares and radio frequency jammers are all parts of defensive aid suites and therefore have to be carried on current target drones. Small sub-targets can also be released. Real-time telemetry from the target drone also provides updates on its position that can be compared with the missile under test. This enables each engagement to be re-played to see the closest point of approach achieved by the missile system. This enables models of the kill probability to be used to decide whether the engagement was ultimately successful.

The contemporaries of the historical target drones are platforms such as the Banshee built by Meggitt. This is used all over the world as a target for air-to-air and surface-to-air engagements. Flying on radar altimeter control, the Banshee can literally skim over the wave-tops. It can also operate up to a maximum ceiling of 7,000 metres (23,000 feet). It is launched from a catapult with recovery by parachute. It can carry a variety of payloads to simulate contemporary countermeasures such as flares and chaff, and can operate over a range of 100 kilometres (60 miles). The Banshee is able to fly an entire mission automatically.

Flying alongside the Banshee at the United Kingdom test facilities at Aberporth is the Mirach 100/X target drone developed by SELEX Galileo. It can fly at transonic speeds and emulate high-performance threats. It can also carry a wide variety of payloads to simulate contemporary countermeasures. It can operate up to a ceiling of 12,500 metres (41,000 feet). The Mirach 100/X has a flight duration of over 100 minutes and is launched with the assistance of two JATO boosters.

United States target drone development

In the United States the development of target drones had started from simple radio-controlled models. In the 1930s Reginald Denny built on his knowledge of how to build radio-controlled models to develop his first concept for a target drone. This was called the Radioplane-1 (RP-1). It was first demonstrated to the United States army in 1935. Subsequent demonstrations also took place of modified variants named the RP-2, RP-3 and RP-4. The latter tests took place on the eve of war in 1939. The United States navy had also been active in experimenting with radio-controlled aircraft in the 1930s. The result of this work was the development of the Curtiss N2C-2 platform in 1937.

At best the initial reaction of the United States military to this development could be described as lukewarm. Shortly, however, with the catalyst being the onset of another world war, the idea gained acceptance. An initial order for fifty-three RP-4s was placed in 1940. They were to enter service as the OQ-1, a designation indicating that they were a subscale target.

The success of the initial tests with the RP-4/OQ-1 led to a larger order from the United States army in 1941 for the RP-5. This entered service as the OQ-2. It had an endurance of seventy minutes and took off from a conventional runway. It could be recovered using a runway or parachute. Guidance was achieved by radio control using a system built by Bendix. With a wingspan of just over 12 feet (3.73 metres) and weight of 104lb (47.2 kilos) it was quite small. The aircraft had a maximum ceiling of 8,000 feet (2,440 metres) and a maximum speed of 85 mph (74 knots). It was powered by a 6hp two-cylinder two-cycle engine.

The United States navy quickly followed suit, ordering the same platform and designating it the TDD-1. Two variants of this were also developed and called the TDD-2 and TDD-3. These were produced in much larger quantities. In the end an astonishing 15,374 of the target drones were built in the Second World War alone. These were variously designated the RP-4 to the RP-18. Many, however, did not get off the drawing board.

A number of events in the Second World War demonstrated the arrival of air power as a significant element of military capability in the maritime domain. The loss of HMS *Prince of Wales* and HMS *Repulse* off the Malayan coast on 10 December 1941 is perhaps one of a number of incidents that showed how air power had developed. This event saw the first major loss of capital ships to air power while the ships were manoeuvring to avoid being attacked.

During the attack on the warships the Japanese Air Force employed both high-level bombers and torpedo-bombers. Eight of the nine twin-engined torpedo-bombers attacked HMS *Prince of Wales*. Attacking at a height of around 50 metres, the bombers each released a single torpedo. Seven were avoided and one struck on the port side aft. The warship began to list heavily to port. The steering gear was also out of control. Another attack by high-level bombers was sufficient to sink the ship. Months later the German battleship *Bismarck* met a similar fate as the result of an attack by torpedoes. At the Battle of the Coral Sea and Midway air power again showed how it now had a decisive edge over naval forces.

The defence of major capital ships of these types to attacks from the air had traditionally relied on their ability to manoeuvre. There was also a legacy from the First World War. Dreadnought battleships had little need for any form of close-in defence. They were generally armed with two 3-inch BL Mk I quick-firing guns that could elevate up to 90 degrees. They had a range of around 11,200 yards at 45 degrees elevation. Against the threat posed by the Zeppelins, these weapons were ineffective. When a Zeppelin appeared at the Battle of Jutland, Royal Navy warships tried to engage it with 12- and

15-inch guns. While the threat from torpedo-carrying aircraft did merit a belated response with an upgrade in the size of ammunition used in the belt-fed quick-firing gun, any further developments petered out once the First World War came to an end.

Naval anti-aircraft gunnery had been shown to have its limitations. As aircraft attacked, the operators had to use simple visual sights to try to maintain their guns on the targets. As the aircraft criss-crossed the sky the chances of a naval battery actually hitting an aircraft were slight. A weapon that typified the problems at the time was the 40mm anti-aircraft gun. It could fire 160 rounds a minute but had to be tracked manually across the sky.

At the end of the Second World War the United States navy and the Royal Navy needed to find ways to improve their naval gunnery. To address this problem they returned to some of the initiatives that had been started in the period between the wars using unmanned aircraft as target drones.

One of the challenges for the development of target drones was how to make them able to fly the kind of high-performance flight envelope associated with fighter jets. As they developed, each generation became more agile and capable than before. The sixth generation of fighter jets is already on the drawing boards of major aerospace companies. They are bound to be increasingly agile.

Flying target drones that can simulate those kinds of flight dynamics is something that requires a platform to have all the same characteristics as a fighter jet. It is therefore perhaps completely understandable that one of the ways to address the problem of developing target drones was to adapt past generations of fighter jets as they went out of service. The latest iteration of that is seeing manned F-16 Fighting Falcons developed into the unmanned QF-16 target drone.

In the United States and the United Kingdom the practice of using former military jets as drones has been well established. An early generation of American unmanned target drones was the PQF-102 which was based on the Convair F-102 Delta Dagger. The conversion of a manned fighter to an unmanned target drone took place under the Full Scale Aerial Target (FSAT) project called Pave Deuce.

The F-102 was an ideal candidate to be converted into a target drone. It was a second-generation fighter jet that formed the main element of the air defence of the United States during the difficult period of the 1950s. Its first flight occurred on 24 October 1953 before entering service in 1956, replacing the Northrop F-89 Scorpion. Over 1,000 of the F-102 Delta Daggers were built. They were retired from active service in 1979. In anticipation of a new life as a target drone, six of the aircraft were converted for remote operation in 1973 and were designated as the QF-102A.

The aim was for the target drone to simulate the performance of the ubiquitous MiG-21. The timing of this conversion was highly prescient as months after the project had started the Egyptians and Syrians faced off against the Israelis in the Yom Kippur War. As the lessons of that conflict began to be digested, a programme was instigated to convert sixty-five F-102s into target drones for the F-4 Phantom and the F-15 when it came into service. These were designated the PQM-102A. The target drone was also used by the United States army to test the first generation of its Patriot ground-to-air missile system. Another variant was also developed: the PQM-102B that allowed for the target drone to be either manned or unmanned, giving greater flexibility in terms of the missions it could perform. These were finally retired from service in 1986.

Overlapping the service life of the PQM-102 variants was the development of the F-100 Super Sabre. This supersonic jet replaced the F-86 Sabre that had flown with such distinction in the Korean War. The F-100 entered service on 27 September 1954 just ahead of the F-102 Delta Dagger and was to fly many missions over Vietnam in a close air support role. It was, however, an aircraft that was dogged by operational problems. In 1967 the airframe underwent a major structural reinforcement programme that was designed to more than double its forecast airframe hours. However, by the end of the Vietnam War 242 F-100s had been lost on combat operations, many to the SA-2 (NATO Code Name Guideline) surface-to-air missile system.

Under the auspices of the FSAT programme nine F-100s were converted by Sperry Flight Systems into target drones. They were designated QF-100. Two (designated the YQF-100) were used for evaluation purposes and were able to be flown by a pilot. Three were converted to become USAF target drones and three modified to perform the same role for the United States army. The last of the nine was a two-seat variant designated the F-100F. Once the proving trials were complete, an order was placed to convert a total of 209 F-102s into target drones. In service in this role the average lifetime of the QF-100 was ten missions.

For take-off two ground-based operators controlled the QF-100 from the end of the runway. Once in the air, handling was passed to another controller in a ground station in the test range area. The manoeuvres conducted during the flight tests were pre-programmed into onboard computer systems. If the QF-100 survived the mission it would be handed back to the two controllers who had supervised the take-off to bring it in to land.

In the latter stages of its career it was used to test the early pre-production variants of the AIM-120 Advanced Medium-Range Air-to-Air Missile (AMRAAM). This was a major development in the field of air-to-air combat as it was a fire-and-forget missile. Today it is in service with over thirty air forces around the world. The AIM-120D variant of the missile is operationally deployed with the F-22 Raptor aircraft. Its predecessor, the AIM-7 Sparrow, had to ride a target designation beam from the radar of the host fighter. It only had a range of 19 kilometres (12 miles). The AIM-7 Sparrow was the first Beyond Visual Range (BVR) missile, entering service after the Korean War.

In order to keep up with the introduction of even more advanced Soviet threats, the United States Air Force turned to the F-106 Delta Dart as the source of its next generation of target drone. In service 342 of these were produced and they served from June 1959 until August 1988. The target drone conversions were finally retired in 1998.

One example of a naval target drone is the BQM-74 Chukar that has been developed by Northrop in the United States. The name originated from an Asian species of partridge that was introduced into America and hunted for sport. The parallels with a target drone are all too apparent.

The BQM-74 has also seen service with the Royal Navy, Italian navy and was used by NATO nations at the multi-national test range facility on Crete. Its first flight was in 1965. It was designed as a high-level target drone that was capable of subsonic operation (up to Mach 0.86). Nearly fifty years later it remains in service, having gone through a series of upgrade programmes with nearly 2,000 having been built.

Its initial development in the 1960s was in part motivated by the threat posed by Soviet naval aviation bombers such as the Tupolev Tu-22M. In the Cold War its mission was to conduct long-range anti-shipping attacks. In the 1970s at the height of the Cold War several Tu-22Ms conducted simulated attacks against United States naval carrier battle groups. The aircraft was typically armed with one or two Raduga Kh-22 anti-shipping missiles. These were armed with either a conventional or nuclear warhead. Once launched, the missile could fly at nearly five times the speed of sound and operate over a range of 320 nautical miles (600 kilometres). It could also be operated in either a high- or low-altitude mode, adding additional complexities to the problems for naval commanders.

Defending against this kind of long-range stand-off missile was a problem. For naval commanders standing air patrols had to be created to shoot down the bomber carrying the weapon before it could be launched.

While subsequent developments of close-in weapon systems did restore some balance to the survivability question, in the early days of the Cold War it was imperative to stop its launch. This had to be done when the Tu-22 was outside the engagement range of the Kh-22. For naval aviators tasked with trying to shoot down the bomber, a target drone such as the BQM-74 provided a useful if not wholly representative test.

Over its nearly fifty years in service the BQM-74 has undergone a number of upgrades. The first delta-wing design proposed for the platform was quickly replaced by a straight wing. Thrust levels available from the turbo-jet engine required that the take-off power be supplemented by a booster. As the threat from Soviet naval aviation developed in the Cold War, the United States navy explored building a target drone that could simulate a vertical take-off and landing aircraft. A prime motivation for this development was the initial introduction into service in 1976 of the Yak-38 (NATO designation Forger). This was to operate off Soviet naval Kiev-class carriers in much the same way as the Harrier aircraft served the Royal Navy. Its top speed was Mach 0.96.

To develop a target drone to simulate the Yak-38, the MQM-74A was adapted and given a new designation of the XBQM-108. Many of the components of the BQM-74 were retained but its power plant was replaced with an engine that was capable of providing rotating vectored thrust. Changes were also understandably made to the undercarriage and flight control systems.

Another UMA that has been in service for several decades is the MQM-107 Streaker. It is used by the United States army and air force as a target drone. For the United States army it is a good target for tests of their surface-to-air missile systems. The United States Air Force uses the MQM-107 to test the effectiveness of its mainstream air-to-air missiles such as the AIM-9 Sidewinder and the AIM-120 AMRAAM. To date nearly 2,500 of these target drones have been built. It has also appeared in a range of variants. Currently it is also in use in a number of air forces around the world.

Soviet target drones
Soviet developments in unmanned aircraft lagged behind those occurring in America, the United Kingdom and Germany. The upheavals after the revolution and the intensity of Russian involvement in the Second World War had left their mark on the Soviet research and development base. When the very existence of a country is at stake, any programme that does not directly contribute to the main war effort is a distraction.

Once the war was over resources could be diverted to new projects. Initially advances in jet fighter technologies preoccupied the main aircraft design bureaus. New research facilities were also created to help develop transport aircraft and helicopters. Many of the names that remain famous today, such as Antonov and Kamov, saw their genesis during this period. Such was the need to play catch-up with developments in the west that there was not a great deal of funds left for the development of unmanned aircraft. Rapid developments in missile technology, however, did create conditions where target drones would need to be developed.

Being ever pragmatic in their approach to the development of new capabilities, the Soviets initially sourced their need for target drones by adapting former manned aircraft into their new role. Aircraft such as the MiG-15 had an additional letter M added to their designation to indicate they were a *mishen* (target). The early variants were manually flown into a test area before the pilot handed over control of the aircraft to a ground station and ejected. If the target drone was to survive the engagement it would be destroyed by remote control from the ground. This approach, however, was not sustainable and the requirement for a low-cost target drone was developed. The first variant of this was the La-17 (Izdeliye 201) which emerged from the Lavochkin design bureau.

The La-17 design was quite simplistic. It was designed to be launched from a large bomber acting as a 'mother ship'. The aircraft first selected to act in this role was the Tupolev Tu-2. It had been mass-produced in the Second World War and was therefore available in large numbers. However, mating the La-17 to the Tu-2 proved problematic. The plan to fly the La-17 from the Tu-2 was quickly abandoned in favour of attaching it to the Tupolev Tu-4 heavy bomber, a derivative of the Boeing B-29.

The design of the La-17 involved a series of compromises. Weight requirements ruled out any means by which the UMA could be recovered if it failed to be destroyed by a fighter in the course of an engagement. The La-17 could therefore fly for a maximum of forty minutes before it would belly-flop onto the ground and onto the engine. Despite this somewhat simplistic approach, photographs show La-17s with several markings on their tails indicating that they have been used more than once.

In the first flight tests a problem with engine thrust became apparent. The top speed of the Tu-4 was insufficient to prevent the La-17 diving after it had been released. The RD-900 engine simply could not generate enough lift at this speed as the air flow into the engine chamber was insufficient. Ramjet engines are simply unable to move on the ground without external power.

They rely on forward motion to compress air into the combustion chamber and work most efficiently at speeds around Mach 3.

Once the La-17 had been dropped from the Tu-4, control often took around ninety seconds to be established. At a speed of around 850 kilometres per hour (528 mph) the La-17 was able to make the kind of manoeuvres required to make the task of shooting it down representative of a current threat. These problems caused the programme to be suspended for a short period of time while modifications were made. After successful flight tests by ten La-17 drones in a variety of roles it entered service with the Soviet Air Force where in various advanced configurations it would remain in service for nearly thirty years.

In contrast to the pulse-jet engine flown on the V-1, the design team opted for a ramjet. This was a far from ideal choice. Fuel consumption rates limited the length of time the La-17 could fly. This gave it a higher overall speed of 900 kilometres per hour (560 mph) and an operating ceiling of 10,000 metres (32,810 feet). This configuration of the target drone allowed it to partially mimic the performance of the MiG-17.

The MiG-17's maximum speed was slightly higher at 1,145 kilometres per hour (710 mph) but its service ceiling was much higher at 16,600 metres (54,450 feet). The ramjet configuration was a sensible compromise that meant the cost of the La-17 could be held down. However, with the introduction of the Soviet Union's first supersonic jet, the MiG-19, into service, the performance of the La-17 soon fell well below that of the fighter jet it was supposed to be simulating. This was not the only problem.

The size of the drone itself posed a problem for contemporary Russian radar system technologies. Its radar cross-section was quite small and needed to be enhanced so the ground-based radar systems could guide the chasing aircraft to a point where they gained visual contact with the target. To enhance the radar signature the target drone could be fitted with a number of Luneburg lenses on the wings and tailplane. These devices increased the radar cross-section of the La-17 by an order of magnitude, allowing it to simulate contemporary threats from the English Electric Canberra or American medium-range bombers such as the B-47 Stratojet.

Despite achieving its initial design goals, other issues were emerging that led the design team to look at developing a new generation of target drone. This was to become the La-17M (Izdeliye 203). It was to be ground-launched from a platform based on a KS-19 anti-aircraft gun mount. This would allow the La-17 to overcome the restrictions of being launched from the Tu-4 which also limited the numbers that could be fired in a salvo. Film taken at the time shows several examples of La-17s being fired in a salvo to create a high-density threat environment. It also had a new radio system and autopilot installed. However, the target drone was hampered by its short range and primitive guidance system.

To launch the La-17M two additional PRD-98 solid-fuel Rocket-Assisted Take-Off (RATO) engines were mounted on either side of the main engine. These had a burn time of between 1.6 and 3.1 seconds. Combined with the main engine running at idle, this generated enough thrust to accelerate the La-17M to more than 300 kilometres per hour (186 mph). Two seconds after launch the main engine was commanded to full power. The boosters were jettisoned after five seconds, at which point the La-17M transitioned to level flight.

After some debate in the design team over the exact configuration of the power plant, a decision was made to use the Mikulin RD-9BK turbojet engine used in the MiG-19. It could produce 19.1 kN (4,300 pounds) of thrust. This was to double the power that was available, although it actually marginally reduced the maximum speed that the target drone could achieve. However, its service ceiling increased dramatically and its flight time increased from forty to sixty minutes.

It was at around this time that rapid developments in missile technologies started to create an increasingly hostile environment for manned aircraft. The ultimate demonstration of this was the shooting down of the U-2 carrying Gary Powers on a reconnaissance mission in Soviet airspace in 1960. Reconnaissance, however, was not a military capability that could be easily given up. The La-17 provided a platform from which a new generation of UMA could be developed that could fly hazardous reconnaissance missions. Yet it was far from being an ideal baseline from which to work.

Chinese target drone developments

Research work in China into UMA had started in the late 1950s. The first developments saw flight control systems developed that could automatically take off and land the Antonov An-2 and 11-28 transport aircraft. Its first target drone was called the Chang Kong-1 (CK-1), a derivative of the Soviet La-17. Its first flight occurred in October 1969, after it emerged from research work at the Nanjing University of Aeronautics and Astronautics. Showing how slow development was in those days in China, the CK-1 did not enter service until March 1977. The one innovation the Chinese introduced into the CK-1 was a rolling take-off from a three-wheeled support trolley (dolly). This was because early versions of the CK-1 relied solely on the turbojet for take-off and that meant speed took time to build up.

The CK-1 was a re-engineered version of the La-17 which had been supplied by the Soviets in the late 1950s. The team leader of this work was General Zhao Xu who is known as the 'Father of Chinese UAV'. At this time China was still learning lessons from its involvement in the Korean War and the aftermath of the first Taiwan Straits Crisis from 1954–55.

Despite its increasing isolation, one of the simple facts that China could not ignore was the development of the missile age. It needed a target drone in order to verify its first generation of air-to-air missiles. Its payloads consisted of packages that could enhance its signature in the radar and electro-optical parts of the electro-magnetic spectrum. Due to the perennial problems the Chinese have suffered with the developments of their power plants for such vehicles, the CK-1 was powered by a WP-6 turbojet that could produce 24.5 kN of thrust. The CK-1 had a range of around 600 kilometres (372 miles) and an endurance of around an hour.

A number of variants of the CK-1 were also developed. The CK-1B was a low-altitude drone, the CK-1C was capable of performing a range of difficult manoeuvres and the CK-1E was a very low-flying target drone. These variants of the CK-1 remained in service with Chinese military forces at the start of the twenty-first century. The CK-2 target drone research and development work started in the early 1990s. This was a supersonic drone. Other developments in China resulted in test flights of the BA-2 (ASN-2), BA-7 (ASN-7) and BA-9 (ASN-9) target drones, although a lack of detailed reporting emerging from China on their use suggests they were not produced in any great numbers.

In keeping with developments in the west the Chinese also modified a Chinese copy of the MiG-17 fighter (J-5) to act as a target drone. This was designated the Ba-5. Reports separately suggest that over 200 J-6 (MiG-19) jets were also converted by the Shenyang Aircraft Company to act as target drones after their retirement from operational duties. The Chengdu J-7 (MiG-21) has also been used in this role. In each case, life support systems were removed from the aircraft and data links and control systems installed to allow it to be controlled remotely.

At that time China had little immediate need for any form of reconnaissance capability, although the CK-1A variant of the UMA was developed to enable remote radiation monitoring. This was at around the time that China became a nuclear power when it detonated an atomic weapon at its Lop Nur test site on 16 October 1964. The CK-1A was initially deployed into the Lop Nur site in 1978. It brought to an end the rather dangerous missions involving manned sampling of the aftermath of atmospheric testing of nuclear weapons in China.

China had entered a rather introspective phase of its development, despite the occasional flare-ups it had with Taiwan and India. Relations with the Soviet Union were often less than cordial. However, the leadership of the Communist Party ideally wanted to stay out of any more military engagements if at all possible. There were too many things to do at home. Uppermost in their minds was the survival of the Communist Party.

Renewed assertiveness would come later as China's massive economy started to develop. Defence spending increases tracked increasing economic power. A surge in expenditure on UMA technologies was always likely to occur once China invested sufficient time and energy into their development.

After a period of little progress in the 1970s and early 1980s a new set of platforms started to emerge from research institutes. China now has a range of development programmes aimed at equipping it with UMA that can play a wide variety of military roles. One of those is the Tianjin-1 target drone. It provides a test vehicle that resembles the kind of flight profiles used by American cruise missiles such as the Tomahawk. The Tianjin-1 entered service with Chinese military forces in 2005. This was also the point at which it was first on public display.

Summary
Using target drones for naval gunnery practice was arguably the first real-world use of unmanned aviation technologies. Until their introduction, military interest in the field was rather muted. While people could see the potential for a guided bomb, technological limitations restricted its roles. As aircraft became more versatile and manoeuvrable it was natural that navies would need to test their gunnery skills against representative targets. Having target drones that could dive towards a warship provided a realistic threat. The emergence of the threat from missiles had a dramatic impact on the development of target drones. New developments would be able to be classified into one of two classes in a simple taxonomy.

The first branch comprised those target drones that were retired former manned jet fighters. These provided agile, real-world-sized targets that would allow engagements to be simulated with a high degree of realism. The second branch was made up of slow unmanned aircraft that were relatively cheap and expendable. Each branch also benefited from developments in technology that helped miniaturize flight control systems and introduced new forms of digital navigation capability.

In the early days of using both systems each played an important part in the evaluation of new missile systems. As time has moved on there is a notable shift towards the unmanned branch of the taxonomy. Part of this is no doubt determined by cost considerations but another element will be the degree to which the aerodynamics of both the missile and its potential targets are now so well understood that actually conducting engagements against realistic targets is less of a requirement. Computer simulations of both the missile and the fighter jets they target are now conducted at such high degrees of fidelity that many missile flight-test programmes can be conducted virtually. But that does not mean that the epitaph for target drones is about to be written shortly.

What is not so easy to test, even on today's high-speed computer systems, are the interactions that take place in the battle between a missile seeker head and the defensive aid suites that fly on the current generation of combat aircraft. That is something that has to be tested in a range of realistic engagement scenarios using real-world examples. This simple point is also recognized by countries whose arms industries are currently under development.

In Turkey the development of the TURNA target drone started in 1995. It entered service in the Turkish Air Force in 2001. It has a flight duration of around ninety minutes. Already export orders are being signed for the supply of the target drone and its control systems. In creating a target drone, nascent technologies can be built that help foster the conditions that in turn help to build indigenous aviation industries, thereby reducing dependence on overseas suppliers.

In places like Iran, India and South Korea indigenous industries are growing up on the back of home-grown development programmes focused on building new jet trainers. As they provide the airframe and engine technologies on which future national fighter jet projects are built, it is inevitable that they will also need to either build or buy target drones. Once those programmes start there is another inevitable outcome. As those indigenous aviation industries develop, the next generation of UMA to emerge will move from being the hunted to the hunter.

CHAPTER 5

Intelligence Collection and Defence Suppression

In charting the subsequent developments of UMA from being a target drone to being able to deliver a weapon onto a target it is instructive to take stock of the capabilities of UMA at a point towards the end of the 1950s. In a fairly short period of time some important developments had been made that laid the foundation for steps that would occur once particular technological breakthroughs had occurred.

UMA could fly at a variety of altitudes, present a realistic (albeit radar-enhanced) target that was capable of varying degrees of manoeuvre, and could be reliably controlled from the ground, provided the UMA was in line of sight of the controlling station. Tentative steps had also been taken by the United States army looking into the use of UMA for carrying small camera systems in a tactical role. This was to be the start of an increasingly important role for UMA that was to provide important operational benefits in subsequent military confrontations.

The platform was called the SD-1: in service it went on to be designated the AN/USD-1. It had been adapted from the OQ-19 Radio-Controlled Aerial Target (RCAT). Improvements in the control and recovery systems had also been made to remove some of the parachute recovery problems that had been associated with the OQ-19. Nearly 1,500 were built between 1959 and 1966. It was the first UMA configured to operate in a surveillance role. As such it was the pathfinder for many of the tactical UMA that were later deployed into Iraq and Afghanistan. Its operating manual defines its mission as 'to perform aerial photographic reconnaissance in conjunction with a mobile radar tracking unit.'

The AN/USD-1 was launched from a small trolley using two jet-assisted take-off bottles. Launch angles of up to 12° could be set. The AN/MPQ-29 X-band tracking radar had a theoretical maximum tracking range of 92 kilometres (57 miles). For each 'drone platoon' that operated the AN/USD-1, two of these radars provided tracking data that allowed decisions to be made as to where to fly the platform to gain the best target coverage. However, field tests revealed that in practice a range of only 45 kilometres (27 miles) could be achieved.

This severely limited the tactical application of the platform. Flight control to the AN/USD-1 was provided by an Ultra High Frequency (UHF) radio that operated between 406-420 MHz. This served as the source of commands to the ailerons and elevators that manoeuvred the platform which could stay airborne for around thirty minutes and fly at a speed of 160 knots. The total payload weight that could be carried was 60lb (27.22 kilos).

Two forms of mission were identified for the AN/USD-1. These were called 'pre-planned' and 'immediate'. For 'immediate' operations any previous mission plan that had been derived for the day to suit the pre-planned tasking was overridden by a request through the command chain.

The AN/USD-1 could be configured to carry either a KA-20A daylight camera or a KA-39A infrared night camera. The KA-20A was mounted in the forward section of the UMA just behind the engine. It produced film in a 9 x 9-inch format that was capable of providing ninety-five exposures during daylight operations. The number of photoflashes restricted nighttime operations to ten exposures. In daylight the AN/USD-1 could operate at altitudes from 1,000 to 5,000 feet. At nighttime the altitude range was reduced to 1,000 to 2,000 feet. The minimum altitude of the UMA was 400 feet. Gyros were also built into the platform to provide the kind of pitch and roll stabilization required for photographic work.

The overall operational concept for the USD-1 was hampered by the absence of technology to download the imagery in real time. It could only provide what was termed at the time 'general target locations in the battle area'. The turnaround from point of intelligence collection to the point at which imagery could be exploited was far too long for tactical applications. A mobile photographic darkroom (ES-29) was part of the ground facilities operated by the support team. Downlinks of video from television cameras were also considered but at the time they were simply not of the quality required for military applications.

To move UMA on to another level of applications from this point at the end of the 1950s required developments in the following areas:

• Navigation of the UMA was always a problem in the early years. Radar tracking provided an update of the location of the platform, allowing its trajectory over the target area to be controlled.

Autonomous operations, where the UMA could decide on the basis of other inputs where to fly to cover the target area, were simply not available. It would take the introduction of Global Positioning Systems (GPS) technologies in the early 1980s to overcome the problems with navigation systems. None of the precursors to GPS were accurate enough.

• For the UMA to move out beyond line of sight some form of over-the-horizon capability needed to be introduced into the radio-control systems. At the time High Frequency (HF) communications relied on Morse-code signalling. Signal strength and reliability could vary significantly during the day. Bandwidth was also very limited in terms of the data rates that could be transmitted over HF channels. Implementing antennas on the UMA to receive long-wave frequencies was also a challenge. It would take the development of digital satellite communications in the mid to late 1970s and reductions in antenna sizes before UMA could really aspire to move beyond the line-of-sight restrictions that existed in ground-to-air communications.

• Technological developments were needed to reduce the size of camera systems and their associated power supplies. Until the UMA could carry a significant payload they were not going to break out of their established roles. Where UMA were to be used to collect intelligence the issue of how to recover the pictures needed to be solved. Either the UMA had to be recoverable and land without suffering damage or it had to be able to eject a cartridge or part of the payload that could be recovered.

• The duration of the UMA mission. While flying on a test range for an hour was considered just about acceptable, to really advance UMA as viable platforms for intelligence collection missions they simply had to be capable of flying for longer periods of time. This meant that engines needed to undergo significant developments.

Decoy developments
In the 1960s progress in all areas was slow. As developments in SAM systems continued, the United States Air Force did investigate the potential for UMA to act as decoys. This was part of an overall approach to using electronic warfare and direct attack methods using radar homing missiles to help suppress the operations of enemy defence systems. The various elements of these missions were brought together under a general heading of the

74

Suppression of Enemy Air Defence (SEAD). This was an umbrella covering anti-radiation missiles that would home in on radar systems that were emitting, as well as electronic warfare activities designed to degrade the operations of the sensor systems employed by an adversary.

The aim in the case of a cheap decoy platform would have been to create a plethora of targets in the hope that it would increase the survivability rate of any bomber force trying to penetrate an enemy's airspace. Research programmes started in the middle of the 1950s produced a number of initial designs that included the XSM-73 Goose. This was a long-range, jet-powered decoy missile.

The first decoy to enter service was the GAM-72 Quail. Twenty-four test missiles were developed before the GAM-72A became operational. In total 592 of these were built and delivered. It was designed to be carried in the bomb bay of a B-52. Up to eight could be carried in the bomb bay, although a more usual configuration was to carry two. In June 1963 the GAM-72A was re-designated the ADM-20B. The UMA was capable of carrying a payload of up to 100lb (45 kilos). This was used to house electronic countermeasures such as chaff, a radar repeater or an infrared burner designed to simulate the exhaust from a jet engine. It was the forerunner of the ADM-141 Tactical Air-Launched Decoy (TALD) system that is in service today.

It entered service with the United States Air Force on 13 September 1960 and was carried aloft on its first alert mission on 1 January 1962. Ten months later it was to participate in the full-scale strategic bomber alert that was mounted during the Cuban Missile Crisis on 26 October when 80 per cent of the Strategic Air Command (SAC) was either in the air or on the ground, held at the highest alert level. On that day one-eighth of the 1,436 bombers available to General Curtis LeMay were on airborne alert, many carrying the Quail.

At its peak in 1963 the total inventory held by the United States Air Force was 492. From this point on the total steadily declined until only 354 remained active in 1977. The UMA had an operational range of between 661 kilometres (357 miles) and 716 kilometres (445 miles) and a flight ceiling of 15,200 metres (50,000 feet). Duration varied according to operating altitude but was typically just under an hour. The flight profile, which was pre-programmed on the ground prior to take-off, allowed two turns and one increase in speed for each mission.

Rapid advances in Soviet radar technology made the task of mimicking a strategic bomber almost impossible. Later developments in electronic warfare were to change that dynamic. This was the classic operational analysis problem. By carrying UMA on some of the bomber force, and by implication removing their bomb load, could the weight of attacks on an enemy be increased? As ever, in such mathematical analysis the outcomes depend to a huge extent on the assumptions made. By 1971 the United States Air Force considered that 'the Quail was only slightly better than nothing.' The Quail, however, can be said to have been one of the precursors of today's cruise missiles.

First steps
Another important development was taking place at the time. In 1959 Ryan Aeronautical carried out a study to look at how it might extend the range of its Firebee drones and convert them into a reconnaissance platform. The launch of the Sputnik 1 satellite had come as a complete surprise to the Americans and although its capabilities were little more than acting as a radio beacon, it heralded developments that would surely follow. Space-based sensor systems are now a routine part of the daily intelligence collection apparatus operated by the Americans, French, Russians, Israelis and Chinese. Even Iran has launched a satellite equipped with a simple imaging capability.

What was urgently needed at the time was a capability to overfly Soviet territory and collect data on their military capabilities. The launch of Sputnik 1 also showed just how far Soviet ballistic missile technology had developed. Assessments emerging from the Pentagon painted a bleak picture. The lack of intelligence led to some huge mistakes in the analysis of the Soviet missile and bomber capabilities. The so-called 'missile gap' and the 'bomber gap' became the talk of all Washington. These concerns, of course, were underlined when Yuri Gagarin made his brief entry into space on 12 April 1961.

Arguably in this climate of fear decisions that were taken led to the arms race. Neither side could be sure of what the others were doing. Intelligence collection assets such as the U-2 were still in their infancy. The sheer scale of the intelligence collection problem was beyond even their emerging capabilities. In that febrile atmosphere it was inevitable that a safety-first approach would be taken. Build weapons first, ask questions of the intelligence analysts later.

This is the background against which Ryan Aeronautical looked into the development of the Firebee. The first mission profile involved flying south from a launch point in the Barents Sea over the Soviet Union to a recovery in Turkey. In April 1960 Ryan Aeronautical presented their findings to the United States Air Force. The timing was impeccable. A U-2 spy plane carrying Gary Powers had just been shot down over the Soviet Union. Two months later a Boeing RB-47 reconnaissance aircraft was shot down by a MiG-19 fighter in international airspace on the border of the Soviet Union near Murmansk. Two of the crew survived and were held in Moscow's notorious Lubyanka prison.

This event was one of many that occurred during the Cold War. In the decade of the 1950s seventy-five US navy and Air Force crews lost their lives in ten reconnaissance missions. For the Americans, the cost of flying manned missions of this type was becoming increasingly high. The first recorded events involving the shooting down of United States Air Force planes by Soviet air defence systems occurred over Yugoslavia on 9 August 1946. Ten days later an almost identical situation occurred. In both cases, transport aircraft had been the target.

In 1952 the United States Air Force lost two RB-29 reconnaissance aircraft near the disputed Kuril Islands north of Hokkaido in Japan. This was to become an area that was in the front line of intelligence collection efforts in the immediate aftermath of the Second World War. Over the next four years a further four United States Air Force aircraft on reconnaissance duties were also lost in the same area.

In one of the most notorious incidents a Lockheed EC-121 Super Constellation SIGINT aircraft was shot down by North Korean fighters in April 1969. The incident occurred in international airspace and was roundly condemned. Thirty-one crew members aboard the aircraft were killed in the incident. It was a small but significant moment. It was to lead a number of teams in the United States to start thinking about how UMA could perform such difficult missions. It also eliminated the potential for embarrassing show trials of captured airmen being trailed on the media.

The United States was not the only country to lose manned aircraft on reconnaissance missions. The Taiwanese were given a number of U-2 platforms in the 1960s to monitor military activities on mainland China. Between 1962 and 1969 they lost six of these aircraft. Clearly the trends all showed that conducting manned reconnaissance missions over areas protected by the first post-Second World War generation of air defence systems was an increasingly risky occupation. This gave impetus to those seeking to find a role for unmanned platforms. One such organization was Ryan Aeronautical.

Their first proposal had envisaged the development of the Model 136 (Red Wagon). Early on, the design team made a decision to forgo the use of an undercarriage. This saved weight and freed up volume for different payloads. Another important initial step that was to give the later Model 147 a great deal of flexibility involved fitting the payload in the nose-cone. Film footage at the time shows how mission-specific equipment was loaded into the front of the platform. This simple idea was to provide a very early form of the kind of modular approach to systems development that is a feature of contemporary approaches to systems. The designers also had to be mindful of issues such as the point where the centre of gravity of the platform would be located. The engine at the back provided a cantilever around the centre of gravity to the payload at the front.

However, the impetus behind the project stalled when President Kennedy arrived in the White House. Ryan Aeronautical received a similar rebuff when it tried to offer another configuration of the Firebee as a SIGINT collection platform under the name Lucy Lee. At this point funding was being directed towards satellite and high-flying manned systems like the Lockheed A-12.

That was to change, however, when the United States National Reconnaissance Office (NRO) awarded a contract to modify four Firebees to convert them into photoreconnaissance platforms. Three months later the Model 147A code-named Fire Fly was delivered. Test flights took place in April 1962 over New Mexico. A camera taken from the U-2 spy plane was installed to provide the sensor system component.

Tests also showed that the Fire Fly had a low radar cross-section that would make it difficult for Soviet radar systems to detect its presence. It did, however, have a problem with the exhaust systems of the engine, leaving a tell-tale contrail in its wake. As the photoreconnaissance Spitfires had shown in the Second World War, this could be just enough to give clues to air defence pilots. When all the guns and ammunition had been stripped from the aircraft to increase its range, anything that compromised its presence was simply bad. Modifications to the Fire Fly ensured that the contrail problem was removed.

Cuban catalyst

The Cuban Missile Crisis was to provide a catalyst for the further development of the UMA in general. The vulnerability of the U-2 to the SA-2 had already been shown in the incident involving Gary Powers. This was the point at which the Soviet SA-2 ground-to-air missile system had quite literally shot to fame. The SA-2 is also attributed with the title of achieving the first combat kill of an aircraft by a missile when it shot down Taiwanese Martin RB-57D Canberra over China in 1959.

During the crisis two other U-2s came under attack. In one, Major David

Anderson was shot down and killed while overflying Cuba. To avoid a repeat incident the Model 147As were authorized to be used to maintain the ability to watch the Soviet build-up on the island. This decision was overturned by the enigmatic General Curtis LeMay. He had other plans for the Model 147As. U-2 overflights were quickly resumed.

This was the period just before the dawn of the digital age. Semiconductor technology with all the benefits it promised was still in the earliest stage of development. Those first attempts at creating integrated circuits also had serious power problems. Many would run at very high temperatures, requiring specific cooling measures to dissipate the heat they generated. Power consumption was also a major problem. In the 1970s these issues began to be resolved and the first generation of integrated circuits was produced. They provided the basis of the first generation of digital communications systems. Developments in radar systems technologies quickly followed.

In parallel with these developments another major military capability was also making rapid advances. The surface-to-air missile system started to show just how dangerous it could be during the Vietnam War. The Soviet S-75 Dvina missile (NATO Code Name SA-2 Guideline) was a weapon that was particularly feared. With the United States developing its long-range bomber forces in the 1950s, it was inevitable that the Soviets would look to SAM technologies as a component of an improved air defence system.

Their development, however, was also to be a catalyst for a new form of mission for UMA. This was where UMA would be flown in a configuration to disrupt the operation of air defence systems. The mission was called defence suppression and was to herald the onset of a new and more aggressive form of electronic warfare.

Accelerating rapidly to a speed of Mach 3.5, the SA-2 initially had a range of 30 kilometres (19 miles). In 1959 an improved variant of the missile (SA-2B) started to appear that could engage targets out to 34 kilometres (21 miles) and up to an altitude of 30,000 metres (98,000 feet). Further upgrades of the missile system occurred as lessons emerged from the Vietnam War and the Six-Day War between the Arab nations and Israel in 1967.

Vietnam
The first SA-2 sites started to appear in North Vietnam in 1965 around the North Vietnamese capital and its major port Haiphong. Permission to bomb the targets was not forthcoming from the United States Secretary of Defense over concerns that Russian operators might be killed. On 24 July 1965 the first American combat loss to an SA-2 occurred when a United States Air

Force F-4C Phantom was shot down. It was the first of 110 aircraft that were to be lost to SAM engagements in South-East Asia. Urgent action to counter the threat saw the introduction into service of the F-100F Wild Weasel aircraft. It carried radar homing and warning equipment capable of detecting the emissions from the SA-2 fire control radar (NATO Code Name Fan Song).

In the course of its involvement in the war the United States military flew over 5 million combat sorties and lost 2,251 aircraft. Of that total 1,737 were lost to hostile action with the F-4 Phantom taking a particularly heavy toll. This was a rate of four losses per 10,000 sorties. In the Korean War this was twenty and in the Second World War the figure was ninety-seven. The trend was, and remains, downward.

The Vietnam War was to prove a pivotal point in the development of UMA and their applications in conflict. In the coming years twenty-eight variants of the Model 147 Firebee would be developed and fly over Vietnam. Each would have its own subtle adaptations and mission drivers. One variant (Model 147NC) carried propaganda leaflets. It was a time when arguably UMA underwent a fundamental change in the nature of their missions. They were no longer 'cannon fodder' for missile tests.

The Model 147 UMA was to be the platform on which the fundamental developments in UMA technology would be made that would dramatically change its role. In 1999 when Ryan Aeronautical was purchased by the American aerospace company Northrop Grumman all the history of these developments would lay the foundation of contemporary UMA platforms such as the RQ-4 Global Hawk. It can trace its ancestry to the important work carried out on the development of the Model 147 in support of the military endeavours in Vietnam and ongoing monitoring of North Korea and China.

At a time when American intervention in Vietnam was only just beginning, President Johnson had concerns that the Chinese may be drawn into the war, mirroring their involvement in Korea. The president authorized Lightning Bugs to be based at the Kadena Air Force base in Okinawa to fly reconnaissance missions over Southern China. The classified operation for this was called BLUE SPRINGS. Five missions were flown over China until early September 1964. The success of Operation BLUE SPRINGS was mixed. The Lightning Bugs had reliability issues which created doubt over their effectiveness as a platform for intelligence collection.

Despite these reservations, operations were transferred to Bien Hoa Air Base in South Vietnam in October 1964. In the coming three months a total of twenty missions was flown using the Model 147B. On 15 November 1964 one was destroyed by the Chinese. This was the first in what was to be a series of losses that would eventually provide the Chinese with sufficient material to reverse-engineer the Model 147B and help create their first indigenous target drone. Three of the Model 147B platforms were put on public display to embarrass the American government. Associated media cover also made the most of the opportunities, praising the success of the Chinese air defence systems. In the United States, in the absence of any captured air-crew being paraded in front of the cameras, the press virtually ignored the story.

Development of the Firebee continued. A Model 147G was produced. Its power plant could generate 8.5 kN of thrust, an improvement over the 7.56 kN of its predecessor. The fuselage was also extended. This was delivered to the United States Air Force in July 1965. A Model 147G flew the first mission for the type over Vietnam in July 1965. The Model 147B was phased out of operations in December of the same year. A low-altitude variant of the platform was called the Model 147J. It had operational issues around terrain avoidance and was also a more demanding environment for the airframe. The Model 147J was ready for operational service in March 1966 and had a new camera system installed.

Paradoxically the low-level missions also improved the survivability of the Firebee as the chances to engage them were limited. Recovery was carried out by a helicopter snagging the parachute from the Model 147 in mid-air. Despite sounding complicated, it was highly successful with 2,655 completed recoveries out of a total of 2,745 missions. In 1966 the Firebee platforms completed 105 missions over North Vietnam and China.

The Vietnam War also saw the first deployment of a UMA to collect SIGINT. Out of a total of seventy-seven missions flown in 1965, three flew with a special payload designed to listen to the electronic warfare environment. The UMA could fly into areas that were simply too hazardous to operate manned platforms. To collect SIGINT that could be exploited required some missions to go into areas where it was likely they would be successfully engaged.

The range of a number of the key radar systems involved meant that to get a good understanding of its operating parameters a SIGINT platform had to get up within line of sight of the radar. If those radar systems were normally deployed in a point defence role, it created a real problem for anyone wanting to gain intelligence on how they worked. Flying UMA into the areas where they would be tracked and possibly engaged by the radar system was essential, especially if the operation of the proximity-fusing arming signal was to be understood. In essence these were unmanned 'kamikaze' missions flown against specific radar systems in order to get up close and personal with the way all facets of the SAM system worked.

The three specially-modified UMA were deployed to Vietnam under Operation UNITED EFFORT. Overheating caused all three of the SIGINT packages to fail on their first missions. After modifications in the United States they were returned to Vietnam. The UMA involved was called the Model 147E, a variant of the Ryan Firebee target drone that eventually became known as the Lightning Bug. It was also equipped with an active radar-enhancing device to ensure the Fan Song radar would detect the platform. On 13 February 1966 a Model 147E variant of the Firebee made history when it detected the command link signal from the Fan Song E radar system used to control the SA-2 before it was destroyed. Arguably this was a pivotal moment in the Vietnam War.

Once the operation of the Fan Song E was understood, measures could be developed to reduce its operational effectiveness. There were several ways in which this might be achieved. For example, one approach would be to try to jam the radar system. That often required a significant amount of power and airborne platforms had their limitations in this respect. Another more subtle approach was to manipulate the operation of the radar itself or the command guidance link to the missile in flight. Each alternative had its challenges.

Having processed the intelligence information collected from that mission, the United States Air Force could set about deciding how to degrade the operation of the two uplink channels used to control the missile. The link itself was found to be unencrypted. The designers had presumably reckoned on not having to bother to encode the link as the chances of an enemy getting close enough to interfere with the command link seemed unlikely. That is not a view that SAM missile designers would take today.

On the Fan Song E command link United States intelligence analysts found four quite distinct commands were transmitted to the missile in flight. These were the K1 and K2 waveforms which commanded the missile to climb/dive or turn right/left. The K3 waveform armed the proximity fuse and the K4 programmed the proximity-fuse delay on the warhead depending upon the engagement geometry. In operation the Fan Song E radar tracked the target and the transponder beacon on the SA-2 and continuously developed an optimal trajectory for an intercept.

Two control laws were also at the heart of the algorithm calculating the flight path. The *Treokh Tochek* or 'three point' control law was based on calculating the line of sight from the radar to the target. This had an important weakness that enabled the missile to be defeated by conducting high-G manoeuvres. But for slow transport aircraft this algorithm would provide a deadly solution. The alternative was called the *Polavinoye Spravleniye* (half correction) technique. This was more sophisticated and was to be used against highly manoeuvrable targets such as fighter jets. The selection of which algorithm to use had a huge impact upon the effectiveness of the SA-2. Warsaw Pact operators and their Vietnamese colleagues proved the more adept at using the missile than other users in the Middle East.

Commentators have suggested that this single mission provided the justification for the development of the entire Model 147 programme. It led to the development of a radar warning receiver called the AN/APR-26 that was to be fitted to United States aircraft operating over Vietnam. It warned the pilot when the command signal became active, indicating that a launch of an SA-2 was imminent. Another development was based on the idea of jamming the Fan Song radar to deceive it. Signals from the Fan Song would be received, slightly delayed and then re-broadcast to the receiver. This fooled the radar system into believing the target was in a different location. This equipment was known as the AN/ALQ-51 Shoe Horn jamming system and it operated in E-Band. The equipment was also carried on the EA-6A Intruder aircraft. In service, however, doubts were to emerge as to the precise effectiveness of the Shoe Horn equipment.

Alongside this specialist mission the United States Air Force was still conducting low-level flights with the Model 147J variant of the Firebee. Its attrition rate was high, so in order to sustain the operational tempo some of the Model 147G platforms were converted to the Model 147J configuration. The Model 147H was used to complement the low-level operations of the Model 147J. It had a more powerful J69-T-41A engine fitted that gave it the ability to fly up to altitudes of 19,800 metres (65,000 feet). While this was on the edge of the SA-2 MEZ (Missile Engagement Zone), it was thought the stealthy characteristics of the UMA would reduce the range over which the Fan Song radar could acquire and track the platform. Time would show that was not the case.

At the time stealth technologies were in their infancy. A special coating of paint applied to the jet intake provided some reduction in the radar cross-section but that was insufficient to reduce the threat. To help improve the survivability of the platform an RWR (Radar Warning Receiver) was added to the payload. This would provide a warning when it was being illuminated by an enemy radar system. A pre-programmed 30° turn would then occur in an attempt to throw off any engagement by an enemy missile. With missile systems agility improving all the time, this was not a fight a UMA could win.

What was needed was a more radical approach. Another box was added to the payload of the Model 147H. This was a jamming system called River Bouncer. It provided a signal that tried to disrupt the operation of the Fan Song radar. These measures did have a positive effect on the survivability of the UMA.

A subsequent refinement of the platform resulted in the development of the Model 147T which could fly up to 23,000 metres (75,000 feet). This entered service in April 1969. The Model 147H was phased out of operations in September 1972. A few months earlier high-altitude flights by the Model 147T over North Vietnam were stopped. The risks to the platform were simply too high. However, as a result of the loss of the EC-121 and all its crew in April 1969 along the border with North Korea, a new variant of the Model 147T appeared. This was the Model 147TE, known as Combat Dawn.

This UMA was the first of its type. The payload was dedicated to the SIGINT mission. It was able to operate at altitudes up to 21,336 metres (70,000 feet). This gave it improved slant range to look deep into a target country to collect SIGINT information. Intelligence material that was collected was downloaded in real time to the controlling ground station. This provided the basis for similar line-of-sight downloading of raw intelligence material on a range of other platforms. Control over the platform could either be accomplished from a ground station or the DC-130 launch aircraft. Recovery of the platform was by helicopter snare.

Initial tests of the Combat Dawn system involved it being launched from a DC-130 aircraft. Once declared operational, Combat Dawn flew twenty-two missions over North Korea and accumulated 61.5 hours in the air. Its attrition rate saw one Combat Dawn lost every ten and a half missions. With the addition of external fuel tanks the UMA was able to increase its mission duration to up to eight hours. Plans were put in place for the development of a twelve-hour mission on a variant of the platform known as the 147TL. Combat Dawn became the primary SIGINT sensor platform used by the United States, flying over 500 missions between its introduction into service in 1970 through to 1975.

It was also around this time that the United States Air Force and the Royal Air Force accepted that high-flying manned bomber missions over the Soviet Union using their strategic bomber force were no longer feasible. If the Royal Air Force's V-Force was going anywhere near the Soviet air defence systems it was going to have to do it at very low level, around 80 metres (250 feet).

These developments of the Model 147H and the Model 147T raise an interesting issue over the cost benefit equation being applied to UMA at this point. As target drones, the issue for designers was all about simplicity. After all, if a missile fired at the target drone was armed there was a possibility that it might actually destroy it. The view had to be that target drones were ultimately disposable objects. However, as the first major revolution in UMA capabilities occurred, what became valuable was the intelligence information that it had collected on its mission. Therefore the priority was to recover the UMA and re-use it on another mission.

Survivability now became the order of the day. If that meant additional electronic equipment such as RWR and jamming systems had to be deployed as part of the payload, then that was accepted as long as the chances the UMA would return back to base were improved. With Soviet radar and technical developments, the high-altitude reconnaissance mission in what was an increasingly hostile (non-permissive) environment that cost-benefit equation meant that new payloads would have to be introduced on the platform. From a survivability viewpoint high-altitude missions were simply no longer viable. Until imagery could be delivered digitally in real time via satellite communication links, other means of intelligence collection had to be found.

Current UMA, such as the Predator and Reaper, operate in permissive environments. Periodic claims by senior Taliban spokespeople claiming that UMA have been shot down are often opportunistic statements allied to crashes caused by operational defects. This may not be the case in the future if a state-on-state war were to break out. The issues over survivability would then re-surface.

In the latter days of America's involvement in Vietnam the Model 147 was to see some final iteration in its design. The Model 147S was a variant whose wing structure was changed to overcome a number of limitations associated with the Model 147B. At low level they did not generate enough lift. They also did not give the platform sufficient manoeuvrability. The wing span was reduced to the original 4 metres flown on the first generation of Firebees.

The Model 147S was also equipped with a new camera system that replaced the dual-configuration camera carried in the Model 147J. The new camera was able to obtain 30cm resolution along a strip 96 kilometres (60 miles) in length. On good days that resolution could even be halved. The new platform went into service as the Model 147SA in December 1967. It was, however, going to suffer a high rate of attrition as it tried to image targets over the heavily-defended areas of Hanoi and Haiphong. Its operating altitude, and hence the viewing geometry of the sensor system, was adjusted in order to let the platform fly at 150 metres (500 feet) instead of the initial configuration three times higher.

The backbone of the UMA sorties over Vietnam was to be the next iteration of the Firebee. This was the Model 147SC. It flew around half the total missions that eventually took place over Vietnam. This was the variant of the Model 147 that was produced in the greatest numbers. It incorporated an improved Doppler navigation system and digital flight controls that helped to improve flight accuracy. It was known as the Buffalo Hunter by its air force crews and went into service in January 1969.

The Model 147SRE was a variant of the Firebee that was dedicated to nighttime reconnaissance. It was equipped with an infrared strobe that provided illumination of the ground beneath the platform. Its imagery was recorded on infrared film. The material collected, however, proved to be difficult to interpret. A Model 147SC/TV also was developed to provide real-time television imagery that was relayed to the DC-130 launch aircraft. However, image quality at the time was low. Improvements were also made to the navigation system with the Model 147SDL including a Loran receiving system. This was an American development of the Gee radio navigation system pioneered by the Royal Air Force in the Second World War for bombing raids over Germany.

Naval variants were also developed that used a RATO booster. These were designated the Model 147SK and entered service in 1969. After take-off the UMA would initially be flown by an operator located in a Grumman E-2A Hawkeye AEW aircraft operating from one of the United States aircraft carriers operating in the area. They flew the UMA to a designated check-point where the control was handed off to the internal flight systems. Recovery was made by helicopter. In total several dozen flights were made by the United States navy under the operational name of Belfry Express. The reasons behind the use of this remain unclear. The last sortie was flown in May 1970 after it had only been in use for barely a year. It had been a tentative initial start for the navy in UMA operations.

In 1968 a total of 340 missions was flown by Model 147 variants. This was over three times the sortie rate achieved in the previous year and more than the total flight operations achieved over Vietnam up until that point. This was a major point in the war when the North Vietnamese launched the Tet (year of the monkey) Offensive on 30 January. This saw coordinated attacks launched across South Vietnam in an audacious bid to win the war. Thirty-six of forty-four provincial capitals were attacked. Fighting was particularly heavy around the United States combat base at Khe Sanh. While it is difficult to establish precisely where the Model 147 UMA were flown during the year, it seems more than coincidental that the sortie rate in 1968 occurred at the time of the Tet Offensive.

In subsequent years the sortie rate for the Firebee increased in 1969 to 437 before levelling off for the next two years. In 1972 at the time of the Operation LINEBACKER bombing raids against strategic targets in North Vietnam the highest recorded rate of sorties occurred at 570. By 1973 this had reduced again down to 444.

Other improvements also occurred in flight control systems technology, such as the gyro systems. The Model 147SB also included the capability to climb and descend the platform between pre-set altitudes to introduce a more random element into its flight path. Slowly but surely, developments in sensor and flight control technologies at this time were laying the baseline for future generations of UMA. In the crucible of war the UMA had taken its first steps from being a target drone used in peacetime to validate the performance of missile systems to being an active capability in a hostile environment.

Over the duration of the American involvement in Vietnam 1,016 Lightning Bugs were to fly a total of 3,435 missions. In the course of flight operations 578 were lost either to enemy action or as a result of an accident. The vast majority of the total missions flown were reconnaissance missions. It became the workhorse of operations in environments that were considered to be becoming increasingly dangerous. In total twenty-three variants of the Lightning Bugs were used in the intervention in Vietnam. One-seventh of them were shot down by ground defences. Some of the UMA attained almost heroic status as they caused Vietnamese pilots to crash while trying to shoot them down. One Lightning Bug was even awarded the status of 'ace' for being involved in five Vietnamese fighters being downed. One flew sixty-eight missions before it too was shot down on 25 September 1974.

The end of the military involvement was not to herald the end of the Firebee. While being at war had created the dynamic for quite specific developments in UMA technologies, as America re-trenched into its post-Vietnam introspection the main focus shifted back towards the Cold War and operations in Europe. Over Vietnam a variant of the Firebee (Model 147NC) had been fitted with chaff dispensers as well as active jamming systems. These were to fly escort missions alongside manned bombers and lay down a carpet of chaff that would disrupt the operations of enemy radar systems.

In Iraq in 2003 Firebees flew their last combat missions performing this role. In total five Firebee platforms were involved in the opening salvos of the Second Gulf War. Only one DC-130 remained equipped to control the Firebee and it was unserviceable on the opening night of the air campaign. Two were therefore launched by RATO. On the second night three were launched from the DC-130. There was no attempt to bring back the Firebees for another mission. They were sent on their way until they ran out of fuel. The wreckage of the platforms was shown on Iraqi television with an accompanying narrative that suggested manned aircraft had been shot down.

After the Vietnam War the SEAD mission was downplayed in new developments of UMA. Image intelligence became the mission priority for UMA in the west and in the Soviet Union and China. The SIGINT mission was not a priority for UMA as far as many states were concerned. That mission would be carried out by manned aircraft, a viewpoint that still applies today. The SEAD role for UMA seemed destined to have been a brief one. That is, unless you lived in Israel.

Arab-Israeli conflicts

The Six-Day War had left indelible scars on the Arab nations involved. Sheer weight of numbers did not count for much when an enemy decided to pre-empt an attack and destroy a large part of the Egyptian air power on the ground. While in the following six days a great deal of hard fighting had to be done, the single blow to the Egyptian military machine was decisive. Six years later in October 1973 the tables were completely turned as out of the blue Egypt and Syria decided to attack Israel on the Day of Atonement. The sheer audacity of this attack ensured that Israel would be forced onto the defensive. Its survival as a nation-state depended on a massive air-lift mounted by the United States.

Historians still debate the degree to which Israeli intelligence failed to provide the detailed warning of the attack until literally hours before the outbreak of hostilities. Israel's intelligence service, the Mossad, had gained an international reputation for being well-connected in senior levels of the Egyptian high command. How an attack could be launched without some word of the plan being leaked baffled the senior people in charge of the organization. However, the difficulties faced by the Israelis in those first few days were not just down to a failure to see a massive military build-up masquerading as an exercise for what it really was, a prelude to invasion.

From the end of the Six-Day War until the start of the Yom Kippur War, the Egyptians and Israelis had been involved in a long-running test of each other's military prowess during what has become known as the War of

Attrition. President Nasser had announced its start on 8 March 1969. It started a period of increasing military engagement between both sides that would culminate in the Yom Kippur War. Frequent small-scale operations were undertaken by each side that aimed to test the defences and state of readiness of one another. Sometimes these small-scale operations would initiate a more robust retaliation.

A familiar pattern emerged that would characterize Israeli military operations until the present day. When Egypt bombarded Israeli positions the IAF would retaliate against tactical and strategic targets. This was part of an approach adopted by the Israelis that ensured an 'asymmetrical response' to any aggressive moves by Egypt. The aim was to ratchet up the cost to Egypt of conducting such attacks in an attempt to deter them from any more military adventures.

Operation RHODES was one mission carried out during this period that was to highlight Israeli concerns about the small but growing network of Egyptian radar systems. Launched on 22 January 1970, it resulted in the capture of a radar system at the Egyptian garrison on the island of Shadwan. This is a barren rocky island around 30 kilometres (20 miles) to the south-west of the Egyptian city of Sharm el-Sheikh. The island is a place well-known to many scuba divers who have visited the northern Red Sea. The successful outcome of the mission was to have an unexpected consequence.

Within days Egyptian President Gamal Abdel Nasser had made a clandestine visit to Moscow where he successfully persuaded the Kremlin to supply his country with advanced weapon systems. The balance of power between Egypt and Israel was about to shift decisively in favour of the Egyptians. Within days Soviet transport aircraft carrying all the building-blocks of a new air defence system began arriving at various air bases in Egypt.

In the coming months these building-blocks were to become the largest air defence system deployed in history. The Egyptians had twenty mobile SA-6 SAM systems. These would range forward in the Yom Kippur War, providing a defensive umbrella over the leading edges of their ground forces in the Sinai Desert. Behind the mobile units there were seventy SA-2s and sixty-five SA-3s at fixed locations supported by over 3,000 SA-7s for point defence duties. The SA-6s, SA-7s and ZSU-23-4s were responsible for the loss of fifty-three IAF A-4 Skyhawks and thirty-three F-4 Phantoms. These losses were to leave an indelible mark on the IAF.

The Yom Kippur War provided a testing ground for Soviet and American technologies. Across the Suez Canal the unthinkable military action that was never to occur across the East and West German border was played out in daily exchanges. Military analysts from the Soviet Union and NATO looked on with increasing interest at the patterns of military activity that were emerging. It was to shape defence policy in NATO for many years to come and lay the baseline in military capability with which NATO would conduct its campaigns in the Balkans and over Iraq in 1991. It was also to be the catalyst for the development in Israel of its own UMA. This was called the Mastiff and was built by Tadiran Electronic Systems. The Mastiff could fly for up to seven hours and operate up to a service ceiling of 4,480 metres (14,700 feet) and could carry a 22lb (10-kilo) load over a range of between 30 and 50 kilometres (18 to 31 miles).

From the western viewpoint this was an ideal opportunity to see how the key elements of the Soviet air defence system, especially its mobile capabilities such as the ZSU-23-4 gun-dish, operated. If the Cold War had ever become the Third World War, this piece of equipment was a key concern for NATO. Its mode of operation was crude. It would simply fill the sky with 23mm lead objects once it had detected an inbound enemy aircraft. It had entered service with the Soviet military at the start of the 1970s. Over time around one-third of the total number built (6,500) were exported to users in twenty-three countries. As far as its Egyptian and Syrian users were concerned, it was the very latest Soviet air defence system.

The ZSU-23-4 was armed with four 23mm auto-cannons and pre-loaded with 2,000 rounds. It operated at a cyclic rate of close to 1,000 rounds per minute providing a combined rate of 4,000 rounds per minute. Sustained fire would last for thirty seconds before the ammunition belts were exhausted. The rotating turret at the top of the ZSU-23-4 also enabled the guns to be laid onto a low-flying target and track it across the sky.

Finding ways of limiting its effectiveness occupied many intelligence analysts in the early part of the 1970s. Initially the solution was to fly as low as humanly possible. Over the vast reaches of the featureless Sinai Desert that created problems for air-crew. It was a weapon that NATO air-crew also feared. It had replaced the previously largely ineffective ZSU-57-2 gun-dish. The ZSU-23 provided the lower level of the Egyptian air defence system. During the Yom Kippur War IAF pilots flying low enough to avoid being engaged by the SA-6 would literally fly into a wall of lead.

What was crucial for the survival of the air-crew was to find ways of skirting around the engagement envelope of the RPK-2 Tobol radar system that cued the ZSU-23-4. It operated in the J-Band and was capable of detecting aircraft up to 20 kilometres (12 miles) away. In poor weather conditions this range could decrease to as little as 7 kilometres (4.3 miles).

Over the Suez Canal and Golan Heights, however, the weather conditions did not trouble the ZSU-23-4.

As the opening salvos of the Yom Kippur War sounded out across the Syrian border with Israel, IAF jets followed established Israeli doctrine and swooped in over the border to attack massed Syrian tank formations. Due to a lack of suitable intelligence collected over the border area the presence of new air defence systems caught the Israeli pilots totally by surprise. During one raid called Operation MODEL 5 (Hebrew *Doogman* 5) six F-4 Phantom aircraft were lost. The ZSU-23-4 element of the Syrian air defence system had exacted a serious toll.

The radar performance against a target flying at 500 knots (1 mile every six seconds) would ensure in a head-on engagement that the ZSU-23-4 would be able to track the aircraft for a maximum of seventy-two seconds before it was overhead. This would allow the radar system to establish a track on the target and wait for around forty seconds before it started to engage. This allowed the tracking algorithm to refine its understanding of the trajectory of the inbound attacking airplane and compute an optimal point for the ZSU-23-4 to open fire.

For the pilot of the incoming aircraft it was in that window when an RWR needed to warn him of the immediate danger. To do that required that SIGINT operations had been carried out effectively against the threats to ensure the pre-formatted messages loaded into the RWR would work. As their losses mounted, the IAF needed an operational RWR into which it could load the results of analyzing SIGINT data. Understanding how the ZSU-23-4 worked was a particular priority.

Israel's SIGINT capability prior to the Yom Kippur War was almost non-existent. For insights into the operation of the SA-2 missile and its radar they had relied on their close intelligence links with the United States. This was to prove a costly oversight. While the intelligence picture on the SA-2 system was good as a result of operations in Vietnam, many other pieces of Soviet hardware were new. Even the United States was struggling to keep up with the pace of Soviet military developments.

The issues that arose from that reliance on the United States for the SIGINT picture will be ones that the Israelis are unlikely to forget anytime soon. It was obvious after the Yom Kippur War that Israel should also develop its own indigenous SIGINT collection platform. For a nation that could not afford platforms like River Joint or the Nimrod R1, a UMA-based SIGINT platform was an obvious way forward. Ideally the solution should be one that was not expendable.

In the course of the War of Attrition the fixed SA-2 and SA-3 batteries were able to account for a small number of Israeli jets that ventured into their MEZ. These were mainly lost due to the lack of an RWR being installed on the IAF aircraft. The only solution in the short term was to pull back. This reduced the ability of the Israelis to gain access to intelligence over the Egyptian lines and to observe any preparations that were being conducted along the Suez Canal which had become the line separating the two forces at the end of the Six-Day War. Observing the overall build-up and disposition of Egyptian forces was simply not possible. It was to prove a fatal lack in capability.

This was at a time when satellite imagery was still not of sufficient quality to completely replace pictures obtained by manned reconnaissance platforms. So the excellent cooperation with the United States on intelligence matters was not a great source of strategic or operational imagery of particular value. The need to see what was happening on the Egyptian side of the Suez Canal became paramount. An urgent mission was sent to America to look at the status of UMA technologies.

After evaluation of the alternatives, a contract was signed with Teledyne Ryan for the supply of an advanced UMA that could operate at both high and low altitudes. Twelve of these new UMA, known as Firebee, arrived in Israel in July 1971. Within weeks they were operational over the Suez Canal region photographing Soviet SAM sites. However, these missions did not help the Israelis compile a comprehensive picture of the Egyptian military build-up. Israel placed too much faith in its HUMINT sources to provide any clues on geo-strategic developments. This was to prove short-sighted and relatively easy to counter. In December 1971 the squadron operating the Firebee UMA was declared operational. The tasks of the UMA were to conduct photographic missions and also to act as aerial decoys. The Israelis gave these UMA the Hebrew name *Mabat* (Glance).

In parallel with this mission another Israeli delegation had also been visiting the United States to explore purchasing a series of cheaper aerial targets that could also act in the decoy role. A decision was made to purchase twenty-seven of the Northrop Chukar target drones. These arrived in Israel in December 1971 and were given the Hebrew name *Telem* (Furrow).

President Nasser's death in September 1970 saw his successor President Anwar Sadat adopt a different strategy. Instead of a never-ending cycle of attack and retribution, he sought to change the dynamic and lull Israel into a false sense of security. It was his careful planning that led to the success achieved by Egypt in the first few days of the Yom Kippur War.

During this conflict the *Telem* UMA were to receive their baptism of fire. In the course of operations against both Egypt and Syria, twenty-three of the twenty-seven *Telem* were flown. Five were destroyed. By contrast, ten Firebees were lost to enemy gunfire during their nineteen missions. This left only two Firebees operational at the end of the war. A further twenty-four were ordered. It was at this time that preparations also started in Israel to build up their own indigenous capability to develop and deploy the next generation of UMA. It was to lead to the development of advanced systems that are now in use in many different countries around the world.

The *Telem* were flown in small groups of two to four at the leading edge of attacks with the specific aim of drawing Egyptian and Syrian anti-aircraft fire in order to allow the following aircraft to pinpoint its source. With many elements of the anti-aircraft systems – such as the ZSU-23 gun-dish – being mobile, the Israelis had a constant problem in trying to find their latest location. Each group would typically attract between twenty to twenty-five Egyptian rockets. Such was their success that during the Yom Kippur War more *Telem* were flown into Israel as part of Operation NICKEL GRASS.

Following the end of the Yom Kippur War the military situation faced by Israel changed dramatically. Egypt, emboldened by its military success during the campaign, sued for peace. Syria, however, could not accept the loss of the strategically important Golan Heights. For the IAF the problem of how to deal with an advanced air defence system remained. There were lots of possible solutions. One involved the development of the ADM-141 TALD decoy. It would fly as an escort to manned fighters and pose such a credible target that any defence system would expend missiles in the belief that they were real aircraft.

Over the coming years tensions would occasionally flare up between Israel and Syria. In one instance the IAF engaged and destroyed an SA-9 battery operated by Libyan troops near Sidon. For the ever-watchful Israeli media, the headlines proclaimed that the solution to the SAM problem that had so hurt the IAF during the Yom Kippur War had been found. In practice this was a small-scale event that did not really signal any significant rebalancing of the situation in favour of the IAF. This was, however, one small step towards a defining moment that was to come. This occurred over a relatively unknown but highly fertile valley in east Lebanon around 30 kilometres (19 miles) to the east of the capital Beirut. It is called the Bekkar (or Beqaa) Valley.

It was over this rather beautiful part of the Lebanon that the west would see its first major SEAD operation against a contemporary Soviet-supplied air defence system. It was to be a confrontation between the IAF and its Syrian counterpart that would last for two hours. In that brief period of time the Syrian air defence system and its air force were savaged by a combination of innovative military tactics.

Lebanon has been a problem for the Israelis for many years. On several occasions, as a result of what it regards as provocations, Israeli ground forces have crossed the border in an attempt to create a *cordon sanitaire* that would move groups that opposed the existence of Israel, such as Hezbollah, back to a point where their rockets were out of range of the settlements in northern Israel. UMA, operating primarily in their air reconnaissance role, were to play a vital part throughout what could be called a 'phoney war'.

By this time Israel's own industrial base had created its first UMA. It was called the Scout (Hebrew *Xahavan* meaning Oriole). This name was derived from an Old World brightly-coloured bird. The first Scout flight became operational on 21 June 1981 and its entry into service was timely. This was two months after IAF fighters had shot down two Syrian helicopters operating over Lebanon and occurred at another low point in Israeli-Syrian relations. Syrian military units were now operating in Lebanon and creating the early stages of an air defence system. In response to the loss of the two helicopters, Syrian air defence units were quickly mobilized into the area to provide cover for ground troops based there. It was also a matter of weeks after the IAF had conducted the highly successful raid into Iraq that destroyed the Osirak nuclear reactor outside Baghdad in a daring attack known as Operation OPERA.

Around the same time on 14 May 1981 an Israeli Firebee was conducting a routine surveillance mission over Lebanon. A Syrian MiG-21 scrambled to engage the Firebee suddenly stalled, sending it spiralling to the ground. The pilot successfully ejected. Like the V-1 engagements over southern England in 1944, another pilot was to discover that shooting down a UMA was not quite as easy as it might seem.

The situation between Israel and Syria was to come to a head just over a year later on 6 June 1982. Israeli ground forces crossed the border into Lebanon. This was the start of what has come to be known as the First Lebanon War. It was to last just over eleven months. The presence of the Syrian SAM systems in Lebanon restricted IAF operations in the area in support of ground operations by the Israeli army. During this period Israeli Firebee UMA flew three missions over Syrian territory, only one of which returned. Nine sorties were also flown by *Telem* UMA with two crashing after launch.

A critical point in the campaign quickly arrived. It happened as the Israeli army moved on a town in the Lebanon called Jezzine. It is located 22 kilometres (14 miles) from Sidon and 40 kilometres (25 miles) to the south of Beirut. Up to this point the Israeli ground forces had managed to avoid a direct confrontation with the Syrian army. On entry into Ain Zhalta in the late evening of 8 June 1982 the Israeli army came into direct contact with the Syrians.

The IAF Scout UMAs were in the forefront of a range of missions at this time. Their main focus was in the location of Syrian anti-aircraft systems. They were also engaged in battle-damage assessment after air and ground strikes. In one operation an SA-8 battery was successfully located and destroyed in a follow-up air strike. During these activities only a single Scout UMA was lost to ground fire.

This provided the catalyst for launching Operation MOLE CRICKET 19. Mole Cricket was the generic name in the IAF for the SEAD mission. The number 19 signified the number of enemy SAM sites that it was designed to destroy. As the battle unfolded, a Scout UMA spotted an additional five SA-6 batteries being moved into the region by the Syrians. This was interpreted by the Israeli Cabinet as the Syrians being keen at all costs to avoid a major state-on-state war at that moment; otherwise the missiles would have been left at their original location defending the approaches to Damascus.

Political authorization for the commencement of Operation MOLE CRICKET 19 was given just after midday on 9 June 1982. An initial wave of ninety-six F-15 Eagle and F-16 Falcon ground-attack aircraft headed for their targets. During the two-hour mission at least two Mastiff or Scout UMAs were in the air over the designated target areas. Imagery from these areas was routed back to the command centres, enabling the position of any mobile elements of the Syrian air defence systems to be monitored. As a result of this close coordination the IAF was able to destroy seventeen of the nineteen SAM sites. Over eighty Syrian fighter jets were also shot down in the course of the battle. No Israeli jets were lost. In just under a decade, Israel had overcome the weaknesses it experienced during the Yom Kippur War and had now found a way of mastering contemporary Soviet air defence systems. UMA had played an important role in achieving that goal. Arguably it was the point at which they came of age. From that point onwards, UMA would become an integral part of nation-states' military and security capabilities.

For the Soviet Union, the Battle of the Bekkar Valley was a bitter blow to their military and political prestige. Their very best systems and aircraft had proven to be no match for the IAF. Some hard thinking had to be done in the Kremlin. Two years later on 11 March 1985 President Mikhail Gorbachev would become General Secretary of the Communist Party of the Soviet Union. It was a position from which he would lead the Soviet Union on a very different course to the previous ideological conflict with the west.

In 1988 Gorbachev was to define a key element of this new approach when he introduced the concept of *glasnost* that was to define a new era of openness and transparency in the Soviet Union. It was one step along a pathway that would see the world's political landscape dramatically change. The culmination of that was the reintegration of Germany. While that was celebrated in many parts of Europe, other developments in the Balkans and the Middle East were to mean that any respite from the Cold War was to be short-lived.

The new era
Iraq's invasion of Kuwait was in many ways a predictable outcome to a long-standing economic dispute over reparations that existed between the two countries. Once the immediate threat of a further intervention into Saudi Arabia had passed there was time to start to think about what to do next. The prime minister of the United Kingdom, Margaret Thatcher, buoyed by her success nearly a decade earlier in the Falklands War, insisted that the invasion could not stand. President George Bush agreed and started the process of establishing an international coalition of forces that would ultimately liberate Kuwait.

In the First Gulf War the allies arrayed against Saddam Hussein used UMA in quite distinct areas. One of the most successful roles was that played by the ADM-141 TALD. More than 100 were deployed on the first night of the air war to great effect. For many weeks beforehand allied planes had been running up to the border region in the dark of night, only to pull away at the last moment and return to their bases. The aim was to wear down the Iraqi defenders. On the first night of the air war the pattern that had been established changed. The allied aircraft did not turn back. Iraqi radar systems that tried to track the large number of inbound targets found themselves quickly being engaged by anti-radiation missiles. The overall SEAD strategy worked.

By contrast, nearly a decade later NATO operations against Serbian forces in the Balkans SEAD would have a limited effect. The Serbian air defence system was entirely based upon components supplied by the Soviets during the period when Yugoslavia existed as an element of the Warsaw Pact. Its disintegration along ethnic divisions created a series of small wars and stabilization operations in which NATO became embroiled. In many of those small-scale operations UMA played quite a limited tactical role. Weather often added an additional limitation to their operations.

However, as NATO started its air campaign over Serbia and Kosovo it was apparent that many of the techniques that had been tried and tested over Iraq in 1991 would again show their military utility. Given the size and effectiveness of the Serbian air defence system, it appeared that SEAD missions would once again be at the forefront of the air campaign. Or that was the plan. What happened in practice was somewhat different.

The Serbian air defence system was based on literally thousands of Soviet-made SAMs. These were a combination of fixed and mobile units. They were equipped with three battalions of SA-2s, sixteen SA-3s directed by the LOW BLOW radar system, and five SA-6 (NATO Code Name Gainful) regiments. Each of these fielded five batteries of SA-6s. The twenty-five SA-6 batteries were also directed by STRAIGHT FLUSH fire-control radars. These had a maximum range of 55–75 kilometres (34–46 miles) and could engage a target up to 10,000 metres. The SA-6 could engage targets out to a maximum range of 24 kilometres (14 miles). Its mobility enabled the Serbians to surprise NATO fighter jets. It was a system that needed to be respected. In the Yom Kippur War it had proven a deadly adversary, inflicting a heavy toll on Israeli fighter jets.

In the run-up to the conflict the Serbians sought help from Iraqi air defence teams that had experienced NATO SEAD operations in 1991. This allowed the Serbians to create new tactics to deal with NATO SEAD activities. The key principle at the heart of their revised approach was to minimize radar emission times. This negated the potential impact of NATO weapons such as ALARM (Air-Launched Anti-Radar Missile) which had been so effective over Iraq in 1991.

Whereas over Vietnam UMA had performed the SIGINT role, it was Nimrod R1 aircraft from the Royal Air Force that were standing off and watching when the radars were emitting. Radio calls to inbound attack packages were quickly made to warn them of the impending danger. Twelve years later off the Libyan coast the Nimrod R1 aircraft saw their last operational missions conducting similar operations against the Libyan air defence system.

Looking at the outcomes of these various military interventions, one thing is clear. Many defence academies across the world have studied the benefits of reducing the emissions from their radar systems to an absolute minimum.

If the SEAD mission is to survive in such an environment it has to adapt. This is particularly true now that digital technologies have given radar systems new operational flexibility. The plethora of waveform types, polarizations, subtle adjustments to frequencies, power levels and adaptive beam-forming capabilities create a highly agile target to be attacked. This complicates the task of trying to create a set of pre-flight messages that can be loaded into an RWR.

As manned fighter jets such as the F-35 Lightning become increasingly expensive, there is perhaps a case to be made for the reintroduction of the SIGINT role for UMA. However, that is not straightforward. The analysis of SIGINT is not something that is readily automated. Radars have war modes that they try to conceal from potential adversaries. Despite intensive intelligence collection efforts in the run-up to the air war, the aircraft approaching the difficult air defence environment around Baghdad in 1991 could not be totally confident that they knew how the Iraqis would use their air defence radars and SAMs. Escorting strike packages into the target area was a role played by dedicated electronic warfare platforms such as the EA-6B Prowler and the EF-111 Raven aircraft.

Automating that level of capability is not easy. When SIGINT is collected it is often incomplete. Like photographic interpretation, it requires a skilled analyst to understand what has been collected. The process from intelligence collection to new programming information being sent out to the radar warning receivers in operational squadrons can introduce delays that place air-crews at risk. That is why the Nimrod R1 operations off Kosovo and Libya were so important. They showed how an aircraft designed for a strategic intelligence collection mission could be used in a tactical role.

It is possible to speculate that in the future the SEAD mission will return to the UMA. One way this could work is if a number of UMA could fly ahead of a manned platform listening to what will be a dynamic electronic warfare environment. Communicating what they see back to the manned platform would enable any defence suppression equipment carried on the UMA to be activated in the minimum time possible. This kind of mission, however, can hardly be justified in the kind of stabilization operations in countries like Mali where there is no appreciable air defence environment to suppress. It would take an industrial-level war reminiscent of the Second World War, the Korean War or the Vietnam War to justify that kind of capability.

Two possible examples of that which exist today concern the disputes over ownership of the sea bed in the Arctic Circle around a geological formation known as the Lomonosov Ridge and the South China Sea. Should either of these cross the boundary from confrontation into conflict, a SEAD mission would become a vital component of any response. Decisions taken in Australia to upgrade twelve of their F-18 aircraft to act as manned electronic warfare escorts reflect a growing recognition that elements of the SEAD mission still have some military utility.

At present the thinking about the electronic warfare elements of the SEAD mission still emphasizes the man-in-the-loop. However, the lessons from the Vietnam War may yet be recycled. One simple and enduring fact about electronic warfare is that whatever effort is expended in peacetime to gather intelligence on the operation and capabilities of a potential adversary's radar systems, as soon as war breaks out things will be different.

Sending manned aircraft deep into an enemy's territory to collect intelligence on radar systems will be increasingly hazardous. This is where the use of a UMA equipped with SIGINT capabilities makes increasing sense. It is something the Israelis see as a cornerstone of their current military capability. In their position it is easy to see why they believe that a SEAD mission on a UMA is important.

Operations by the IAF over Syria or in the future over Iran would be opposed by the next generation of Russian air defence systems based on the S-400 missile system (NATO Code Name Growler). It is based on two radar systems. The first is the surveillance radar (91N6E) with a range of 600 kilometres (372 miles). This is supported by the multi-functional 92N2E with a target detection range of 400 kilometres (248 miles). Both include a range of digital signal processing capabilities to reduce the impact of noise jammers. The S-400 is based on three different types of missile. They are the 40N6, the 48N6 and the 9M96. Each has a varying range over which it can engage targets. Selection of which missile to fire depends upon the trajectories of the incoming threats. For the United States, its allies and Israel this is a key target for intelligence collection efforts. Suppressing its operational capabilities would be vital if any military interventions were to be contemplated against either state.

While the SEAD mission for UMA has been the subject of some uncertainty, the situation over the reconnaissance mission is clear. That is now firmly written into the ORBAT of military forces around the world. Getting close to near real-time information on the current disposition of enemy forces remains an essential task; never more so now that many countries are scaling back their investment in defence systems. When conflict does arise, the pressure to bring it to a conclusion quickly with the minimum of bloodshed will remain important.

In Mali French forces were supported by a range of unmanned and manned surveillance platforms in their quest to find and fix the location of Islamic militants who threatened to move on the capital city. As the French forces became involved they were faced by two columns of rapidly-moving forces. Air strikes called in as a result of high-quality intelligence provided by UMA changed the dynamic of the situation. Once the militants fell back it was a question of monitoring their retreat and harrying them all the way to their former strongholds in the north-east of the country.

UMA will continue to conduct surveillance operations over the Adrar des Ifoghas Mountains in Mali to monitor any attempts by the Islamists to create new terrorist training camps in the region. They will also monitor the border areas with Algeria and Niger for signs of cross-border activity. Similar UMA activity will also carry on over Afghanistan, Pakistan, Yemen and Somalia against Islamist groups operating in the area. Some will inevitably be armed missions. For the Israelis, UMA have been increasingly integrated into their military operations. The time between sensing a possible threat and it being attacked continues to fall.

Countries such as Pakistan, India and China will continue their developments of UMA. They too have major border issues and internal insurrection problems in remote and difficult to access parts of their countries. In the field of maritime security UMA will also consolidate its role, providing the wide-area surveillance and persistence that satellite-based sensor systems cannot deliver.

In the 1960s China also became aware of a number of overflights of its territory by AQM-34N Firebee UMA on reconnaissance missions. A combination of its limited duration and top speed of 1,140 kilometres per hour (710 mph) meant that the Firebee was never likely to penetrate deep enough into the Chinese hinterland to gather what might be termed valuable strategic intelligence information. America was known to be operating U-2 aircraft out of Pakistan at the time so it is possible that Firebees were deployed to look across the areas where China and India had territorial disputes. Satellite data was simply not of sufficient resolution to enable detailed assessments of the military dispositions of each side at the time. The Firebee was the obvious platform to use and it had successfully operated on overseas deployments.

Several of these platforms were shot down by the Chinese. They wasted little time in analyzing the pieces. A programme to reverse-engineer the Firebee was started and culminated in the development of the BUAA WZ-5 reconnaissance drone. Production started on the WZ-5 at a modest level in

1981. It was China's first indigenous UMA developed for reconnaissance. Initially it was designed to be carried into the air by a TU-4 bomber. On approach to the area of interest the WZ-5 would be released at an altitude of between 4,000 and 5,000 metres (13,123 to 16,404 feet). On ignition of its own propulsion system the WZ-5 would climb to its operational altitude of around 17,500 metres (57,415 feet). It had a range of around 2,500 kilometres (1,550 miles) and could therefore easily overfly Taiwan or parts of India and return.

Since the development of the WZ-5 China has also built several more UMA with an increasing emphasis upon longer range and higher duration missions that would match its growing influence and interest in the South China Sea. The BZK-005 Giant Eagle reconnaissance platform is a long-range, high-altitude UMA developed in China by Harbin Aircraft Industry Group. It is in use by the Chinese navy and was first shown to the public at the Zhuhai International Air Show in 2006. A small number of stealth features are embedded in its design and it also has a satellite communications antenna built into a dome that sits atop the platform. Underneath the platform sits an electro-optical sensor system. The BZK-005 is believed to be capable of flying at a speed of around 170 kilometres per hour (105 mph) and has a service ceiling of 8,000 metres (26,250 feet) with a maximum take-off weight of 1,200 kilos (2,645lb). Its sensor payload weight is also believed to be around 150 kilos (330lb). In August 2011 pictures appeared on various Chinese internet chat forums showing what was believed to be a BZK-005 that had crashed in a field close to Xingtai in Hebei Province which is close to 320 kilometres (198 miles) from the sea. The location of the crash appeared random as no Chinese naval research facilities lie close to the area. It is therefore possible to speculate that the platform was involved in a test flight overland when it crashed. The loss of the BZK-005 may indicate that despite its huge investment in UMA technologies, China still has some way to go before its new platforms achieve a high degree of operational reliability.

In the future the applications to which UMA will be applied will inevitably grow just beyond the defence arena. In Ireland UMA have been used to monitor extensive heathland fires and in the United States they are being used to patrol the long border with Mexico, countering economic migration. FBI sources have already confirmed that UMA have been used on a small number of covert operations in the United States. In the Sudan the United Nations has suggested possible roles for unarmed UMA to support ongoing efforts at stabilization in the wake of Southern Sudan's emergence as a new nation-state. When natural and man-made disasters strike in the future, UMA will almost inevitably be mobilized to help in rescue operations. This will not be the end of their development. Further applications will inevitably arise.

These applications of UMA all use the sensor technologies that began their development in the early 1960s. Over the next forty years those developments would enable UMA to gain greater insights to what was on the ground. In the last two decades of the twentieth century what also changed was the duration of the missions that could be performed by UMA. The wars in the Balkans, Iraq and Afghanistan provided the motivation to push new developments in engine technologies. Advances in electronics also provided the capability to remotely download the pictures collected by the sensor systems and to control the UMA thousands of miles away from the base from which it was flying. The changing international security landscape also provided a pivotal moment when UMA were going to move to another level. After 9/11, UMA moved from being the hunter to being the predator. They were now armed and extremely dangerous.

CHAPTER 6

Predatory Instincts

The result is more than a 'three-bloc war': it is a shifting 'mosaic war' that is difficult for counterinsurgents to envision as a coherent whole.

United States Army Field Manual
3–24 December 2006

The death of an iconic insurgent

The death of the leader of the Pakistani Taliban, Hakimullah Mehsud, on 1 November 2013 was in any sense a major event in the history of attacks conducted by armed UMA. He was by some margin the most senior member of an insurgent group to die in such a manner. While others who have died as a result of armed UMA strikes have also had some profile, such as Anwar al-Awlaki, the leader of Tehreek-e-Taliban Pakistan (TTP) was the first leader of a major terrorist group to die as a result of the actions of the Americans' use of armed UMA.

For the Pakistani Taliban, his loss is a serious but not grievous blow. A new leader, Maulvi Fazlullah, was quickly appointed, even if there remains some uncertainty over his ability to unite what is a fractious group under his leadership. He will, however, lead the attacks in Pakistan from inside Afghanistan. This will further complicate the already difficult relationship between the two countries. A backlash against the Americans based in Afghanistan may also occur.

The timing of the death of Hakimullah Mehsud was interesting as overtures had been made to the Pakistani Taliban to see if they might be drawn into negotiations over some kind of political solution to the problems inside Pakistan. His death was seen to torpedo that initiative, with the president of Afghanistan among those who felt that the death 'took place at an unsuitable time'. Others, as ever, took a different view, suggesting that the death of the TTP leader would not have a significant impact on the pathway to peace. Indeed, some even suggested it may help.

For those inclined to believe the Americans would do anything to prevent Islamabad from reaching an accord with the Taliban, the attack was seen as a deliberate attempt to halt any progress towards a peaceful solution. Others saw the attack as inevitable. Once a man with his reputation came into the cross wires of an armed UMA, the decision to attack the car in which he was moving was not difficult. His past track record as a self-confessed ruthless murderer was sufficient for the attack to be given the go-ahead.

Interestingly, while some Pakistani commentators lamented his passing, others took a different viewpoint. The Sunni Ittehad Council (SIC), a leading Pakistani religious body, released a statement denouncing the title of 'martyr' that had been allocated to Mehsud by some of his followers and by some other notable religious leaders when he died. The chairman of the SIC, Hamid Raza Rizvi, said 'the collective opinion from all (thirty) Muftis said that calling a man responsible for the loss of so many lives a "martyr" went against the teachings of (the) Quran and Sunnah.'

While religious leaders debate the interpretations of Islamic teachings with respect to the death of Hakimullah Mehsud, the insurgency in Pakistan continues. Early reporting provided by *IHS-Jane's* does suggest that the death of Mehsud has had a small but noticeable impact on the tempo of operations being conducted by the Pakistani Taliban. Data accessed on 1 December 2013 showed that in November 2013 only 135 incidents were recorded in the *IHS-Jane's* Joint Terrorism Information Centre (JTIC).

Given the data from which the information was extracted, the overall figure for the number of attacks is likely to increase as reporting is consolidated. However, the figure does contrast somewhat with the data from the previous six months where in May (234), June (173), July (194), August (169), September (207) and October (177) attacks were recorded. Any decline in the rate of attacks conducted by the Pakistani Taliban, however, is likely to be short-lived. It is axiomatic in such situations that the death of a senior leader of a terrorist group rarely leads to the demise of the group. There are simply too many young men, some with even more extreme views than their predecessors, who are ready to step up to take on the leadership role. That does not mean that armed UMA strikes are not worth conducting, as even brief respites can be helpful for the hard-pressed Pakistani government to think of how to adapt and change its tactics given the emerging situation.

The attack on Hakimullah Mehsud came the day after another armed UMA strike had killed at least three people in Miranshah in North Waziristan when a compound and a vehicle located in the local bazaar were attacked. This was the first armed UMA strike for a month, with the previous strike having occurred on 30 September.

It also came two days after one of the top suicide bomb-makers in

Somalia, Ibrahim Ali Abdi, was killed in an armed UMA strike in Somalia that was probably launched from a base in Djibouti or from Arba Minch in southern Ethiopia. This event occurred one week after the attack on the Westgate shopping mall in Nairobi in Kenya. He was killed while driving in a vehicle in the Middle Juba region of Somalia.

This attack followed a failed attempt by United States Navy SEALs to capture a senior member of Al Shabab, the terrorist group operating in Somalia linked to Al Qaeda, known as Ikrima or Abdukadir Mohamed Abdukadir. The location of the attempt by the United States SEALs was Barawe in southern Somalia which had been named as a major training centre for suicide bombers in a United Nations report published earlier in 2013.

Evolving capabilities

The evolution of the roles of UMA in the Cold War followed an obvious trajectory. Their role in collecting intelligence information in non-permissive military environments was always bound to increase, despite the enduring capabilities of manned platforms such as the SR-71 and the U-2. The problem with the SR-71 was that in flying as fast as it did, it required refuelling on a regular basis. Its advantage was that its imaging sensor system could see objects at 12 inches resolution from 80,000 feet. While the SR-71 could outrun enemy missiles, the U-2 could not. It now has to operate in permissive environments. However, even that is limited by human endurance. A typical U-2 mission lasts for twelve hours. The physiological demands on the human body are high. UMA of course suffer no such restrictions.

According to the International Institute for Strategic Studies (IISS) in 2013 eleven different countries were operating fifty-six different types of UMA. Not all of these are armed. Countries that actively employ UMA include India, the United Kingdom, Italy, Turkey, France, Germany and Israel. France and Germany are two countries that lack armed UMA. In an attempt to catch up with developments in America and Israel, a consortium of aviation companies announced the formation of a 'drone users club' on 19 November 2013.

One reason behind this was that France had to reply upon American UMA during its operation against insurgents in Mali in January 2013. For the French this would not have been a position that they would have enjoyed. They like to have their own indigenous capability, meaning that they can operate where and when they decide it is right in their national interests. In the summer the German Defence Minister Thomas de Maizière signalled an intention to work with France to develop a new generation of armed UMA.

A European equivalent to the Reaper and Predator capabilities developed by the United States will soon be moving from the drawing board into production.

In April 2010 India became the latest country to formally announce that it would be using UMA to monitor the activities of communist insurgents belonging to the Communist Party of India (Maoist) (CPI-M) operating in the border region between the states of Chhattisgarh, Orissa and Andhra Pradesh.

The analysis also suggests that 807 UMA are in active service with 678 being operated by the United States in the form of eighteen variants. Around 100 of these are thought to be the MQ-1B Predator operated by the United States Air Force (USAF). The more recent Reaper UMA which is armed has seventy-three platforms in service with the USAF. They are not the only operator of these two specific platforms, with the Air National Guard reported to own forty-two MQ-1B Predators and fourteen MQ-9 Reapers. The United Kingdom and Italy also operate the MQ-9 Reaper.

A number of the operational MQ-9 Reaper platforms are also being upgraded with modifications that extend their flight endurance up to thirty-five hours. A strengthened undercarriage is also being fitted that increases the maximum take-off weight by 12 per cent. One other proposed enhancement sees the wing span increased from 66 feet to 88 feet, further increasing flight durations in some configurations of the UMA to forty-two hours. These are important developments enhancing the operational flexibility of the existing fleet of armed UMA that are being carried out in parallel with the development of the next generation.

Royal Air Force UMA operations in Afghanistan
The United Kingdom flew its small fleet of MQ-9 Reaper UMA almost continuously from the point they entered service in 2008. By August 2010 the Ministry of Defence was happy to reveal that it had spent more than £500 million on purchasing and leasing UMA to support operations in Iraq and Afghanistan. Half of this total had been invested into the MQ-9 Reapers. The United Kingdom MoD also admitted in November 2010 that 15 per cent of its missions had involved some form of kinetic action. This amounted to 293 Hellfire air-to-surface missiles and fifty-two Paveway bombs being fired and dropped. UMA in the service of the Royal Air Force had formally moved from being hunted to being the hunter. They had gained predatory instincts.

In September 2012 the United Kingdom Ministry of Defence noted that the five MQ-9 Reapers had flown for 39,628 hours and fired 334 laser-guided

TDD-2 target drone being launched from a US navy warship. More than 9,400 of these were manufactured. (*United States Department of Defense*)

OQ-3 target drone being tested at El Paso in Texas in 1941. It could reach a speed of just over 100 miles per hour. It was first flown in December 1943 and was an upgraded variant of the OQ-2 with a sturdier steel tube fuselage. This was the United States army variant of the naval TDD-2. (*Righter Family Archives*)

The Radioplane Company developed the OQ-7 which had improved performance over the OQ-3. It had a slightly swept-back wing which enabled the platform to fly at 112 miles per hour. However, it was never produced in volume. (*Righter Family Archives*)

OQ-14 target drone had a higher performance than the OQ-3. It had started out as Radioplane's RP-8 but was of heavier construction and fitted with a more powerful engine that was rated at 16kW (22 hp). It was designated the OQ-14 in service with the Army Air Force and TDD-4 in the United States navy. (*Northrop Corporation*)

Reginald Denny, one of the key players behind the development of drones in the United States who first proposed a radio-controlled aircraft as a target in 1934. The designated TDD-2 adopted by the United States navy stands for Target Drone Denny-2. (The American *magazine*)

B52-D in flight launching a Quail decoy. The idea was to use these to saturate the Soviet air defence system in the event of war. (*United States Air Force*)

The Lavochkin La-17 target drone was the first to enter service with the Soviet Union in the 1950s. The first variations were air-launched but the illustrated variant is a ground-launched drone. It could achieve a maximum speed of 560 miles per hour. (*Source unknown*)

A view of a C-130 Hercules drone control aircraft carrying BQM-34S Firebee target drones mounted on its wing pylons in 1975. (*United States Department of Defense*)

A Firebee drone leaves its launch pad during air-to-air training exercise 'William Tell' 1982. (*United States Department of Defense*)

BQM-74E Chukar target drone equipped with jet-assisted target packs takes off from the flight deck of the USNS *Amelia Earhart* (T-AKE-6) for an air gunnery exercise in the South China Sea in July 2010 during Cooperation Afloat Readiness and Training (CARAT) Singapore 2010. (*United States Navy*)

A standard ER/SM-2 (RIM-67) surface-to-air missile launched from a VLS (Vertical Launch System) intercepts a BQM-34A target drone in 1980. (*United States Department of Defense*)

BQM-74 Chukar target drone over the South China Sea in June 2011. The platform is capable of speeds up to Mach 0.86 and can fly at altitudes of up to 40,000 feet. (*United States Department of Defense*)

BQM-167A Ryan Firebee target drone in flight on Exercise 'Combat Archer'. Its first flight was in 1955 and it is one of the most widely-used target drones ever built. (*United States Department of Defense*)

The MQM-107E Streaker target drone flying in 2004. More than 2,000 of this drone have been built and it is primarily used by the United States Air Force and army for testing and training. (*United States Department of Defense*)

The QF-4 target drone is an unmanned Phantom aircraft used to test the manoeuvrability of air-to-air missile systems in realistic combat situations. It is the latest in a long line of aircraft that have been adapted to be flown remotely. The QF-16 is the next generation of the target drone capability. (*United States Department of Defense*)

Royal Air Force Reaper unmanned aircraft being operated in Afghanistan. (*Royal Air Force*)

Royal Air Force Reaper airborne over Afghanistan, armed with four Hellfire missiles and two Paveway bombs. (*Royal Air Force*)

An artist's impression of the next generation of UMA. (*United States Department of Defense*)

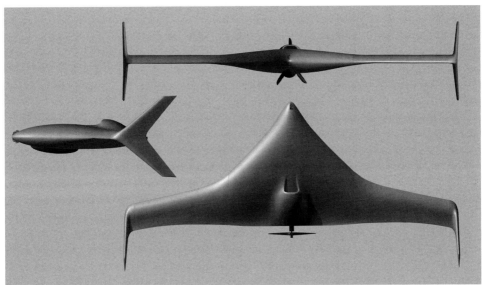

The United States navy X-47 on a test flight. Note the stowed arrester hook. (*United States Navy*)

The United States navy flight-test X-47 vehicle on the ground. (*United States Navy*)

Hellfire missiles and bombs at targets in Afghanistan. Fully loaded with Hellfire missiles and laser-guided bombs, the MQ-9 Reaper can stay airborne for up to eighteen hours. When flown without weapons, this increases to an operational duration of close to thirty hours. The United Kingdom Reapers have a range of 3,700 miles (5,900 kilometres) and a maximum airspeed of 250 knots. They can operate up to an altitude of 50,000 feet (15,300 kilometres).

Since Royal Air Force operations started with UMA in 2006, over 100,000 hours of persistent surveillance had also been generated up until October 2012. None of the United Kingdom UMA are involved in operations across the border in Pakistan. In October 2012 the United Kingdom announced it was doubling the size of its MQ-9 Reaper fleet based in Afghanistan. At the same time the decision to repatriate the control centre for the Royal Air Force UMA from its original base at Creech Air Force Base in the Nevada desert to RAF Waddington was confirmed.

While these figures were disclosed by the United Kingdom Ministry of Defence, this was not accompanied by any greater sense of transparency over the day-to-day use of the Reaper platforms in Afghanistan, despite attempts to obtain information using Freedom of Information Act requests. These culminated in a legal battle that saw the campaigners fail to get the Information Commissioner to tell the Ministry of Defence to release more insights. The argument put forward by the MoD that 'disclosure of the requested information was involving "risk to life and limb"' won the day.

While the campaigners involved expressed their disappointment at the outcome, they vowed to carry on trying to force the Ministry of Defence to be more transparent about the operations of the Reaper platforms in Afghanistan. Meanwhile, the MoD steadfastly maintains its position that it does not target terrorists outside Afghanistan and is not involved in assisting the CIA-led activities in Pakistan. One thing that did emerge from the discussions, however, was that the Ministry of Defence admitted that one armed UMA strike carried out on 25 March 2011 had killed four civilians and wounded two others.

The MQ-9 Reaper had been introduced into service in the Royal Air Force under what is called an Urgent Operational Requirement (UOR). While owning UMA had been part of some long-term plans in the MoD that were due to come to fruition some years later, the decision was taken to bring forward their procurement. The threat from IEDs in Afghanistan was one of the drivers for this decision.

As time moves on and the draw-down from Afghanistan appears on the horizon, it is possible that some of the Royal Air Force Reapers may remain in Afghanistan beyond the pull-out of combat forces. These will be needed to support any enduring operations by United Kingdom Special Forces and provide help to Afghan National Security Forces as they try to maintain a non-permissive environment for terrorists in their country. What had been introduced under a UOR has all the hallmarks of being a permanent feature in the future inventory of the Royal Air Force.

Armed drone attacks in Afghanistan
Gaining any deeper insight into the way that armed UMA strikes are used in Afghanistan is difficult. Attributing attacks to American or United Kingdom UMA is not straightforward, given the lack of transparency of the operations. However, reports do emerge in the Afghan press that purport to provide a commentary both on situations where the Taliban claim to have shot down UMA and also where strikes against targets allied with the Taliban occur.

One area in Afghanistan where reporting of armed UMA strikes has been particularly frequent in the Afghan media is Kunar Province. This is an especially difficult area for ground forces to operate in as nearly nine-tenths (86 per cent) of its terrain is classified as mountainous or semi-mountainous. Only one-eighth of the terrain is officially classified as flat. In such a difficult area – one that also shares a border with those areas of Pakistan most associated with terrorist activity – it is understandable why armed UMA strikes are a key weapon in the fight against terrorists operating in the region.

Significantly at the point at which the main combat forces began their withdrawal from Afghanistan an announcement that the Afghan Air Force had been given its first unmanned surveillance drones was made on 11 November 2013. These are clearly not armed. This may signal that the Americans are keen to retain an armed UMA strike capability in Afghanistan post the military pull-out in 2014. This may help to manage the problem of the insurgency in Kunar Province which is proving difficult to overcome. The scale of the problem in the area is illustrated by some examples of armed UMA strikes in the region.

The death of the shadow district governor of Watapur District in Kunar Province on 9 August 2013 illustrates the degree to which senior member of the Taliban are being targeted. His name was Samiollah. On 22 August 2013 reports emerging from the Brolo area of the Marawara district of eastern Kunar Province in Afghanistan suggested that five Taliban insurgents had been killed in an armed UMA strike. On 14 September it was reported in the Islamic Press Agency that twenty-two members of the Taliban had been killed in an armed UMA strike. Four senior commanders of the Taliban were among those killed. This attack also took place in the Kunar Province of Afghanistan in the Khalaq Lam area of the Chapa Dara District. These attacks were followed up on 19 October when five Taliban died in another strike in the same area. On the same day another two members of the Taliban died in a similar armed UMA strike.

Reports emerging days later suggested that local Taliban had killed a civilian named Shingol who they accused of providing intelligence to the Americans. This highlights one of the other impacts of armed UMA strikes. The Taliban often fear that local people are acting as spies and using mobile phones to provide information on the movements of their key leaders, in effect facilitating the strikes. These reprisals do not endear the Taliban to local people who are often randomly accused of being involved.

Another important factor in the use of armed UMA strikes in Afghanistan is the death of individuals that are in the country having travelled from locations such as Western Europe. The story of the death of a radical Islamist who wanted to be known as 'Ahmad' in an armed UMA strike in October 2012 illustrates the disruptive effect such attacks can also have on the wider international security landscape. He is reported to have originated from the town of Setterich near Aachen in North Rhine-Westphalia. He had joined the group known as the Islamic Movement of Uzbekistan (IMU) and had attended a terrorist training camp located along the border between Pakistan and Afghanistan. There he reportedly specialized in the manufacture of remote-controlled bombs.

His death created pressure within Germany from groups opposed to armed UMA strikes for the official prosecutor to investigate the legality of the attack. Their view was that any citizen of a foreign country killed in an armed UMA strike was protected by international legal provisions. This is an argument that has also been trailed in the United States over the death of Anwar al-Awlaki. The problem of what to do about individuals who travel overseas to become involved in international terrorism is a difficult one.

As citizens of their adopted countries they are supposed to enjoy some degree of legal protection from what can be argued as an extrajudicial killing: one carried out beyond the boundaries of national legal frameworks. It is a difficult subject. What rights do people enjoy when they travel overseas to become involved in international terrorism? How do existing international legal arrangements cover their case? It is a problem faced by people from a range of European countries as evidence emerges of individuals from France, Germany, the United Kingdom, Norway, Sweden and The Netherlands – to name a subset – travelling to Somalia, Pakistan and Afghanistan to receive training in terrorism.

Responding to the death in a similar attack of another 20-year-old German citizen named Bünyamin (Benjamin) who originated from Wuppertal, the Federal Prosecutors Office announced in July 2013 that it was ceasing the investigation as his death was covered by the international law on conflicts as the person in question was a member of an armed group operating inside Pakistan. The office noted in their judgement that 'lethal drone attacks are to be seen as war crimes only if the killed person had the status of a civilian protected by humanitarian international law in times of war.'

While this judgement is unlikely to finally stop the arguments over the legality of armed UMA strikes that kill citizens of countries who have travelled to Afghanistan and Pakistan with the aim of becoming involved with international terrorism, it does at least provide a view of the interpretation of international law. It is a view of jurisprudence that the Americans and British will also no doubt be keen to follow.

The next generation
In the United States the next generation of armed UMA is already appearing. The third jet-powered prototype of the Predator C Avenger aircraft entered flight testing in July 2013. The fourth vehicle is slated to be ready by the spring of 2014. This new development of the series of armed UMA will also carry a new sensor suite called SYERS-3 in its payload bay to collect improved multi-spectral resolution imagery. This will provide the Predator C Avenger with even more capability to discriminate targets on the ground. The current armed UMA boast an accuracy of around 9 feet which is very high, even by contemporary weapon systems standards.

Further developments in these fields are driven by the operational necessity to reduce those occasions when the operators make mistakes and civilians die as a result. These accidental engagements can occur because the sensor suite on board the armed UMA cannot provide a wider context. As the operator becomes fixated on a specific target, it is possible to lose a wider situational viewpoint. This has been likened to drinking information through a 'soda-straw'. One solution to that problem is called the Wide Area Persistent Stare (WAPS) sensor system, a version of which called the Autonomous Real-Time Ground Ubiquitous Surveillance Imaging System (ARGUS-IS) is under development. This has sufficient resolution to watch and record in real-time activity over a much wider area. Its applications in civil law enforcement operations in America are obvious. However, signals that UMA are going to be increasingly used in the United States as a toolkit for law enforcement agencies have not been greeted with a great deal of enthusiasm.

The camera on the ARGUS-IS is based on an astonishing 1.8 gigapixel sensor array that can image objects as small as 6 inches on the ground from an altitude of 17,500 feet. The array is built up from 368 imaging sensor systems based upon cellphone technology. Coupled to the camera, advanced software can track the movement of individuals and objects such as cars. This is especially important in an urban context. In rural situations the current trend with armed UMA strikes is to try to wait for those in the cross wires to be in a car and away from other people who might be killed accidentally.

Maritime UMA
The most common UMA in service is the smaller RQ-7A Shadow of which IISS estimates 236 are operational. These are tactical platforms whose role is to help commanders on the ground in convoy or on manoeuvre to be aware of potential hostile forces gathering to launch attacks. In the era of UMA, tactical and operational commanders should no longer be surprised by events and that equally applies in the naval domain. UMA concepts are also rapidly developing for use in the maritime environment. In 2009 Vice Admiral McCullough provided an insight into the United States navy's thinking when he noted it had plans to spend upwards of $6billion on unmanned projects in the coming five years. Not all of this spending would go on UMA. Some would be directed towards Unmanned Surface Vehicle (USV) developments. This, the admiral noted, was part of a broader plan to embed UMA into day-to-day maritime operations.

With frigates and destroyers routinely deployed with helicopter aviation support to extend the ability to look over the horizon in missions designed to counter piracy and drug-smuggling plus actions to curb the spread of weapons of mass destruction, it was natural for maritime UMA designers to at least consider a helicopter-based platform. The MQ-8C is the latest UMA to emerge in this area. It provides a 30 per cent increase in range and a massive 40 per cent uplift in the payload carried by its predecessor, the MQ-8B.

In 2009 the MQ-8B was seen as the natural adjunct to the Littoral Combat Ship (LCS). It would provide the ISTAR element as well as carrying the Coastal Battlefield Reconnaissance and Analysis (COBRA) shallow-water mine detection system. As those developments matured, maritime UMA capabilities have also moved from being the hunter to the predator.

For the moment those are restricted to surface-based threats such as pirates and those involved in smuggling. In time, developments will also no doubt take place using UMA in the anti-submarine warfare role. A UMA deploying a sonar system into the water to listen for the presence of an enemy submarine is not a far-fetched idea. Neither is the thought that one day a UMA could launch a torpedo. Weight restrictions, however, make that problem slightly more difficult to solve. A Hellfire missile on a Predator carries a small warhead. This would not achieve a great deal against a nuclear submarine or a surface combatant. Arming maritime UMA therefore remains in its infancy. However, as the counter-piracy missions in the Indian Ocean have shown, the problems with trying to gain situational awareness over vast tracts of ocean are ideally suited to the current generation of UMA.

Twenty-three RQ-4B Global Hawk platforms are also operated by the USAF. Four of these have been converted to provide test vehicles for the Broad Area Maritime Surveillance (BAMS) mission in support of naval operations. The overall figure produced by the IISS is likely to be an underestimate, given that the information provided by them did not include data supplied by Russia, Turkey and China.

Precise effects
As the technical challenges of staying aloft for lengthy periods of time were overcome, the potential for UMA to provide an 'eye-in-the-sky' over either a static or moving target was always going to be attractive. However, the crucial factor was their ability to deliver military effect with increasing precision. The random nature of the V-1 attacks towards the end of the Second World War had to be replaced by an ability to deliver the desired effect while minimizing civilian casualties.

As the years have passed the tactics used to try to destroy or capture unmanned aircraft have continued to vary in their sophistication. In Afghanistan when Reapers and Predators have crashed, the Taliban have often been quick to claim it was their actions that resulted in the UMA being lost. Much of this is empty rhetoric. In July 2013 a United States Air Force Predator UMA was captured on film crashing in mountainous terrain in the border region between Iraq and Turkey. Kurdish separatist guerrillas operating in the area claim to have shot down the aircraft. Such rhetoric is hard to verify but UMA operating in mountainous regions are generally more vulnerable to ground attack. The unarmed UMA in question was operating over this area as part of a covert operation called NOMAD SHADOW. The UMA had been supplied to the Turkish government by the Americans. This operation had been activated in November 2011 as concerns mounted over the activities of Kurdish separatists hiding in northern Iraq. This purchase of UMA by foreign governments experiencing problems with terrorist groups that may choose to target western interests is unlikely to be the last.

Terrorist attacks in Turkey mounted from the safe haven of northern Iraq were routinely targeting western interests. The deployment of unarmed American UMA into the area provided real-time downlinks of data to Turkish forces pursuing the Kurdish rebels as they sought sanctuary across the border with Iraq. In time these will be phased out as Turkey's own indigenous UMA development programme comes to fruition. It has been announced that these will be armed. In the meantime, unarmed UMA operated by the Turkish forces have been monitoring activity in the border region with Iraq where they operate.

In October 2013 the Turkish newspaper *Zaman* announced that the UMA had witnessed insurgents preparing to leave Turkey as peace talks with the Kurdistan Workers' Party (PKK) have made progress. Perhaps the outlook in the border area between Turkey and Iraq will see an improving security situation. If that welcome development does occur, it will still be important for UMA to monitor the area in the medium term to ensure the PKK have finally pulled back from conducting terrorist attacks inside Turkey.

In Afghanistan the situation is a little different. While the Taliban do have a small remaining stock of what is often referred to as anti-aircraft artillery (or Triple A), these are rarely found in the areas where the UMA operate. Other weapons used by the Taliban simply do not have the range and homing capability needed to engage a UMA, even though they do not fly that quickly.

On 4 December 2011 the Iranians claimed to have seduced one of the Americans' most advanced unmanned surveillance aircraft, the RQ-170 Sentinel, by conducting an attack upon its control links. The Iranians suggested that they had managed to conduct a cyber-attack upon the radio links used to remotely control the aircraft. The aircraft, the Iranians claimed, had been detected in their airspace 225 kilometres (140 miles) from the border with Afghanistan. It was presumed to be on a mission to monitor developments at various Iranian nuclear sites. The Americans were quick to pour scorn on these claims, suggesting that the aircraft had simply suffered a malfunction and crashed.

Months later an unmanned aircraft, thought to have been flown by Hezbollah, flew from the Lebanon over the Mediterranean Sea and over Israel before it was shot down by an F-16 fighter jet. Its target would have appeared to be the Israeli nuclear power facility at Dimona. Analysis of the remains of the platform tied its design to that of an Iranian unmanned aircraft. Presumably this was a reprisal for the American overflight of Iranian nuclear facilities. Such was the slow speed of the unmanned aircraft that the F-16 had to make several passes before it could launch a missile that destroyed the target. As events over southern England had showed in 1944 and 1945, despite being a relatively easy target to attack, the UMA is not necessarily as vulnerable as people might think.

As if to underscore their increasing capabilities in UMA technology, in November 2013 the Iranians unveiled their own indigenous armed UMA which they claim can fly over a range of 1,200 miles (2,000 kilometres) and can remain airborne for up to thirty hours. The Iranians have decided to call the new UMA *Fotros*. It is named after a fallen angel in Shia mythology who was redeemed by Husayn ibn Ali, a symbolic figure in Islam. This new addition to Iran's military capability shows how they are able to defy international sanctions associated with their nuclear programme and still make significant developments in aviation technology.

While the Iranians seek to make some serious publicity from developing their own armed UMA, they still face practical problems with its operation. Without the kind of sophisticated global satellite communication systems that the American and United Kingdom Reapers rely upon to convey their imagery to ground controllers, the Iranian UMA will have to operate in line-of-sight of the ground controllers. This will limit their operational manoeuvre. Iran's increasing concerns over some of the drug-smuggling that goes on along its borders with Pakistan and Afghanistan may be the first place where the new UMA capability will be used. It is interesting to wonder whether a member of one of the cartels involved in smuggling narcotics into Iran may be the first victim of an Iranian UMA strike. Using armed UMA against criminals will cross a rubric that the west has been reluctant to take so far.

New horizons

As the international security environment changed from the ideological clash of the superpowers in the Cold War to a more uncertain landscape where threats hide in the least accessible areas of the world, the utility of the UMA has suddenly blossomed. Their ability to linger over a target area for an extended period of time watching for an opportunity to strike provides a unique capability that is ideally suited to contemporary forms of warfare. Their ability to lie in wait for a fleeting target is a particular quality that has military and political value. However, that is not where the increasing ubiquity of UMA ends.

Applications exist in the military, law enforcement and disaster management arenas for such a capability. Other novel applications of UMA are emerging, such as mapping fields of vines to check for infestations. In simple terms, anything that can be detected by visual, radar or infrared sensors may eventually become an application for UMA. What has been pioneered in the field of remote sensing from space is now applied at even higher resolutions from UMA. They can also stay over the target area for longer; a vital attribute that differentiates UMA from satellites whose re-visit rate poses problems in building up behaviour patterns of individuals on the ground. This makes UMA such as the Global Hawk of particular interest to specific countries; for example, South Korea and Japan. South Korea's focus is understandably on the land border with its petulant neighbour. Japan has a greater interest in maritime applications of UMA.

From countering piracy in the Indian Ocean to tracking drug smugglers across the Gulf of Mexico and spotting economic migrants travelling across the ocean into Australia, the list of ways of using UMA will inevitably grow. In the future many more applications will no doubt arise where cameras need to be placed into risky and dirty situations where humans find it difficult to venture.

Of all these potential ways of using UMA, the wars in Bosnia, Kosovo, Iraq, Afghanistan and the Israeli issue with Hamas and Hezbollah provided the catalyst for greater investment in the technologies and their rapid evolution from being mere hunters to predators. It also reflected a change in the dynamics of warfare. In the wake of the sacrifices made in Vietnam, the Korean War and the two world wars, western political leaders had become increasingly reluctant to adopt the use of force to achieve political ends. The nervousness in America and Western Europe over attacking the Iranian nuclear programme provides one contemporary example. The reticence over launching military operations in Syria provides another.

Yet as the world has changed, the need for measured and proportionate measures to disrupt the activities of international terrorist groups such as Al Qaeda, Hamas and Hezbollah has also increased. The implications of failing to act to prevent future mass casualties such as those that occurred in America on 11 September are simply too profound.

Increasingly, however, as armed UMA strikes have created disquiet, new forms of interventions are required. A new paradigm is being debated. This is one that moves from the palliative approach adopted so far to one based on a prophylactic model. Whereas in the past the aim was to gain relief from the threat posed by Al Qaeda without addressing the condition that created the threat in the first place, the next evolution will try to stop any future evolution of similar groups ever occurring.

For some who regard themselves as pragmatists and realists, such a solution may appear fanciful. Others may take a different viewpoint, arguing that any form of military intervention in foreign states shows a lack of imagination on the part of political leaders; a position that is also increasingly being challenged by western public opinion, as events concerning a possible military intervention have shown. If western public opinion has no appetite for increased military engagements, other than to target individuals who may be planning major terrorist attacks in the United States and Western Europe, the argument of retaining and perhaps even expanding the use of UMA strikes is given further impetus.

Irrespective of which argument has the political ascendancy at the time, one important fact remains clear. All forms of struggle move on. As the public in western societies finds it increasingly hard to justify sending their men to war, other ways of preventing potentially catastrophic forms of terrorism have to be found. In the short term, the pragmatists argue that requires the use of armed UMA.

In the summer of 2013 a number of aspects of the ways in which armed UMA strikes are planned in the United States began to move into the public domain. One aspect of this that has been given a great deal of scrutiny is the ways in which the so-called 'Disposition Matrix' is used to target specific individuals. By any other name the Disposition Matrix would be called a 'kill list'. For the Obama administration that has too many echoes of the deck of cards approach adopted by George W. Bush.

In fact, suggesting that the Disposition Matrix is a kill list in all but name is to miss a vital point. The database, as it is referred to, contains a range of bibliographic details on potential targets, known associates, current standing in organizations etc., as well as possible ways in which these individuals could be brought to justice. The matrix is an attempt to codify the process by which targeting decisions have been taken by the president into a more formal structure. It has its complications in that lawyers debate the legitimacy of any form of what appears to be extrajudicial killing but the Disposition Matrix at least covers all the bases in terms of the ways in which the targets can be pursued.

Towards the end of 2013 a number of examples of the Disposition Matrix at work were observed when the United States SEALS tried to capture Ikrima in southern Somalia. In parallel with that raid another key member of Al Qaeda, Anas al-Liby, was captured in Tripoli. He was wanted for his alleged role in the two attacks on United States embassies in Africa in 1998. These two raids occurred over 3,000 miles apart but within minutes of each other. One was successful and the other failed but it showed the Disposition Matrix in action. One message the Americans were sending was clear: not all leading terrorists will be killed by an armed UMA strike. If they feel that they can capture and bring an individual to justice they will try to do that, with all its obvious benefits in terms of what intelligence information may be reaped from arresting such a person.

Of course, killing or capturing people is all well and good but the question of how to stop people being drawn into terrorism in the first place is a vital one. In the longer term the approach must be to tackle the root causes of why people become radicalized. That does not necessarily require armed intervention in the form of UMA. If that could be minimized, so the argument goes, one of the causes of radicalization of the people living under the threat of armed UMA strikes would be removed. This raises an interesting and important question. Is there any empirical evidence to suggest that armed UMA attacks do actually radicalize people?

Do UMA attacks radicalize people?
Revenge, it is said, is a dish best served cold. The implications are clear: do not act too hastily or with too much military power, or you may suffer the consequences. It is a phrase that many Israelis understand very clearly. It is also a view that is understood in the White House. Retribution for the attack that killed seven members of the CIA in Khost took its time but on 8 March 2010 Hussein al-Yemeni was the latest member of Al Qaeda's leadership to be struck off the targeting list. He was a Yemeni by birth and was among the top two dozen leaders on the American targeting list of Al Qaeda.

He had, according to senior American officials, played a key role in the Christmas attack on the CIA base. The attack that killed al-Yemeni was reported to have been the first conducted in a built-up area. Clearly at the time the Americans were prepared to take risks to ensure that a key intermediary between the Haqqani network in Afghanistan and Pakistan and the core leadership of Al Qaeda in Pakistan was killed. He was also believed to have had important connections into Yemen.

The need for caution in seeking retribution is particularly true in the field of Counter Insurgency (COIN) operations. The action-reaction – or tit-for-tat – cycle that is so often the catalyst for continuing the violence rarely seems to be broken, especially if the societal backdrop for the military operations is a country in which blood feuds and the importance of revenge is written into the core values and belief systems of those involved. In May 2010 there was a particularly devastating attack on extremists hiding in Pakistan. A reported eighteen missiles killed fourteen fighters in North Waziristan and injured a further four people. A wave of protests followed, driven on by extensive media coverage of the attack.

This was the third attack in what appeared to be a high rate of strikes following the failed attempt by an American citizen known as Faisal Shahzad to detonate a car bomb in Times Square in New York. The surge of strikes in North Waziristan following the attempted bombing looked suspiciously like a direct attempt at payback. To many in the media Al Qaeda and its followers were being sent a very clear message: threaten the United States and you will unleash a whirlwind of retribution. Of course, if the Americans do surge their armed UMA strikes they must expect that on occasions this will cause a backlash.

The idea that UMA strikes cause people in Pakistan to become radicalized is quite fashionable in left-wing circles. The centre-left German newspaper *Süeddeutsche Zeitung* published an article on 7 February 2013 written by Nicolas Richter that argued this position. The author was specifically critical of what he saw to be the hypocrisy of President Obama – having won the Nobel Peace Prize – in turning UMA on innocent civilians on whom they had a 'traumatizing effect'.

The lack of transparency or Congressional oversight of the process by which targets are selected is a common position of left-wing commentators who argue the immorality and illegality of UMA strikes. In his article Richter also took issue with a specific comment made by John Brennan, a senior member of the Obama White House team, who sought to compare UMA strikes to a scalpel. Brennan had claimed that UMA strikes were similar to a surgeon 'cutting out a tumour without damaging the healthy tissue'.

Richter's counter was to suggest that those operating the UMA are 'so enraptured with their new tool that they have long since lost the sense for the quiet minimalism of a surgeon.' The implication of his overall comments was that America may well come to rue the day it sought to flout international legal norms to achieve a short-term effect on international terrorism. In time, Richter argued, other countries would develop UMA capabilities and at that point turn them against American interests.

The French commentator Georges Malbrunot writing in *Le Figaro* on 8 October 2013 took up the theme that UMA strikes actually have a limited impact upon international terrorist networks. The article noted that no fewer than twenty-five of Al Qaeda's top leaders have been 'neutralized', including the American of Yemeni extraction Anwar al-Awlaki who was thought to be behind the radicalization of many people who had visited his internet sites and listened to the 2,000+ sermons he had published online. Malbrunot argued that even when UMA strikes are combined with Special Forces operations designed to arrest other key players, the underlying terrorist networks are able to reconstitute themselves and carry on. These, he argued, were insufficient to weaken a terrorist network that has become a 'label' in the long term.

His point, that in effect Al Qaeda's ideology has greater enduring powers than those individuals who offer their lives in the cause defined by Al Qaeda's narrative, is one with which it is difficult to argue. He also raises the point that from a geographic aspect many of those involved simply enjoy a great deal of geographic manoeuvre room in which to avoid the watchful eyes of UMA. Those terrorists aligned with Al Qaeda that withdrew from Mali in February 2013 simply faded away into the ungoverned spaces along its borders with Mauritania, Algeria, Libya and Niger. The fact that the Libyan government admitted on 11 August 2013 that it was under pressure to allow armed UMA to conduct operations over its southern border illustrates the problem of the vast ungoverned areas in which international terrorist groups can still find sanctuary. Malbrunot also noted that a number of recent high-profile prison breaks had further served to boost Al Qaeda's ranks in places like Pakistan and Yemen.

The dynamics of the increasing geographic range of armed UMA operations is clearly something that the United States will have to confront. With the legal and ethical arguments being made about the validity of these attacks being questioned repeatedly in public, the Obama administration started to subtly modify its position in the summer of 2013. Whereas previously armed UMA strikes were seen as preferential to placing boots on the ground, two Special Forces raids, conducted in the wake of the assault on the Westgate shopping mall in Nairobi, provided evidence that the White House was prepared to use other options available in the Disposition Matrix to target specific high-profile individuals.

Indeed, just before that attack in Kenya the United States Secretary of State, John Kerry, had ventured to suggest on a visit to Pakistan that as far as the armed UMA strikes were concerned: 'I think the program will end as we have eliminated most of the threat.' He went on to say: 'The President has a very real timeline and we hope it's going to be very, very soon.' Anyone looking at how the international terrorist network has managed to adapt in the face of the onslaught from armed UMA strikes might suggest that the US

Secretary of State was engaging in some highly wishful thinking. As if to underpin that point, the White House was quick to distance itself from the remarks made by Kerry.

In the kind of complex tribal societies that have formed the backdrop to current operations, local customs, creeds and traditions are important; none more so than the way a family's honour or status can be lost. If this occurs as a result of not avenging an attack it can have profound implications for the individual, their family and their status in society. The sanctions that can be applied are defined in local interpretations of customary law and can include a range of measures, one of which is for the individual or family to be ostracized from society. In close-knit communities where families may be reliant on one another to survive in their harsh living environment, this can be a very difficult position in which to be placed.

These are parts of the complex and shifting 'mosaic war' acknowledged in Field Manual 3-24: the American military's doctrinal statement of the ways of conducting COIN operations. It has led many to suggest that the use of armed UMA strikes against key leaders involved in the insurgency has a very negative effect on the local population. With pictures of the aftermath of attacks rapidly being sent around the world via social media, the problems become particularly acute. The people directly affected by the attacks, it is often reported, are willing to become directly involved in the insurgency to avenge the deaths of their families and friends. The suggestion is that around the world as armed UMA strikes increase, more and more people will be drawn into cooperation with international terrorist groups such as Al Qaeda and Hezbollah.

For many inclined to criticize the operations of UMA in far-off theatres of conflict, the radicalizing effect of armed UMA attacks is a key concern. It is their single reason for calling a halt to the use of armed UMA. They cite protests on the streets and surveys conducted in the areas affected as providing evidence of the impact moment. What they actually do later is a very different matter. As ever, the Latin phrase *fallaces sunt rerum species et hominum spes fallunt* applies: 'Appearances are deceptive and the hope of men is thus deceived.'

Operations by the Predator and Reaper UMA in the campaigns in which they have been used have led to situations where civilians have been caught in the crossfire of war. Understanding precisely the number of civilians killed in such attacks is difficult. Much of the reporting is based on innuendo and rumour. The need for the family members of those killed to find some form of retribution is important in some societies.

That said, the prevailing view on just how many civilians have died in armed UMA strikes appears to be changing. On 31 October 2013 the Pakistani government issued a statement that said only sixty-seven civilians had died in all the armed UMA strikes since 2008. In comparison with some of the estimates produced by other organizations this is an extraordinarily low figure. Many organizations have ventured into the territory of trying to develop methodologies that can provide more accurate reporting of the level of civilian casualties. Some have tried to manipulate figures published in open sources. In a report published by the Columbia Law School in 2012 these methods were sharply criticized and were described as 'potentially highly misleading'.

The Bureau of Investigative Journalism claims that at least 300 civilians died in the same period. Indeed, the statement released by the Pakistani Ministry of Defence goes further, suggesting that no civilians died in armed UMA strikes in 2012 and 2013 while noting that 2,160 militants had been killed in the attacks. The release of these figures brought the inevitable chorus of conspiracy theorists suggesting that the Americans had somehow paid the Pakistanis to release this low figure.

The truth no doubt lies somewhere in-between these extremes of analysis, although one organization unlikely to be specifically on-side with armed UMA strikes, Amnesty International, has noted in a report published in the autumn of 2013 the degree of intimidation from the Taliban faced by relatives of those killed in armed UMA strikes to exaggerate the numbers of those who have died. One specific uncertainty centres on the problems of how the Americans choose to classify those caught up in the attack. Any males in a specific age range are not regarded as innocent bystanders if they happened to be in a room in a compound that was attacked.

Despite these remaining concerns and the continuing opacity of the White House on the armed UMA strikes, there is a discernible trend of reports that are now producing a less sensationalist viewpoint of the numbers of civilians that are actually being killed. It would appear that one of the prime reasons why some commentators like to venture that armed UMA strikes are a cause of radicalization might, in fact, not be quite what it seems.

While every effort is made to use these instruments of war with precision, errors do occur. When they do, depending upon the prevailing situation at the time, they can have a disproportionate and wide-ranging effect. These can have geo-strategic, strategic, operational and tactical consequences across the entire gamut of the so-called instruments of power. Whereas in the past the line between what was a tactical, operational and strategic military activity was clear, today those lines of demarcation are very blurred. Indeed, some may argue that there is little in the way of discrimination at all. Something that happens at the tactical level can have an immediate and quite profound impact on the strategic landscape.

In Appendix A the insurgency in Pakistan is analyzed. Behind the reporting provided in the Appendix is a great deal of analysis work aimed at looking for any detectable links between armed UMA strikes and any indicators that they have had a radicalizing effect on the ground. Empirically it is very difficult to show that claims made in the media about this coupling actually exist. This is a conclusion that has also been derived by an expert team deployed by the RAND Corporation who have studied data drawn from both Pakistan and Afghanistan with a view to trying to see whether UMA strikes are in some way increasing the levels of violence in those two countries.

Their report, published in July 2013, developed and tested a number of hypotheses using data they collected from a number of similar sources to those analyzed in Appendix A. Their conclusion is interesting. The results, the RAND team note, 'lend credence to the argument that drone strikes, whilst unpopular, have bolstered United States counter-terrorism efforts in Pakistan – and cast doubt on claims that drones strikes are militarily ineffective.' They also go on to develop five different measures of militant violence in Pakistan. These are:

- Frequency of attacks
- Lethality of attacks
- Number of attacks on tribal elders
- The number of IED attacks
- The number of suicide attacks.

They then venture that the detailed analysis of their results did not support a hypothesis they had developed suggesting that increased UMA strikes would be associated with increased terrorism. 'On the contrary', they go on to note, 'they support the alternative hypothesis that drone [sic] strikes are associated with decreases in militant violence.' Powerfully they also add: 'We find no evidence in support of the competing hypothesis that drone [sic] strikes increase violence.'

The implication of this is that the UMA attacks are not adding to the factors that already drive some people into radicalization. In drawing this conclusion the RAND team also looked in detail to see if there were any signs of geographic displacement of terrorism as a result of UMA strikes. None was forthcoming. Any careful analysis of the situation on the ground in Pakistan would quickly suggest that the RAND team's conclusions make sense.

The terrorism picture in Pakistan is very complicated and driven by many local factors. In Balochistan, for example, an insurgency exists to try to create a separate state. Issues over the exploitation of natural resources in the area and the benefits that accrue to people in Balochistan have provided a long-standing set of grievances. In Karachi much of the terrorism that occurs on the streets is sectarian in nature, flaring up periodically in a cycle of tit-for-tat killings. The one area where it is possible to hypothesize that armed UMA strikes would be having a radicalizing effect is in the North-West Frontier Province (NWFP) and the Federally Administered Tribal Areas (FATA). However, analysis of the attacks on the ground conducted by insurgent groups active in the areas does not produce any evidence that such a correlation exists.

Paradoxically where the main set of armed UMA strikes occur in South and North Waziristan there are few retaliatory attacks against pockets of Pakistan security forces located on the ground. There is certainly little evidence in the analysis suggesting that a counter-strike occurs within hours or days of an attack. If anything, the figures suggest that the armed UMA strikes have a coercive effect on the terrorists. Analysis of the patterns of terrorist attacks prompts the question: does this mean that the retaliation is geographically displaced?

The answer to that also appears to be that there is little empirical evidence to suggest that this is the case. Appendix C explores this idea of geographic displacement further by looking at the attacks on NATO convoys passing through Pakistan en route to Afghanistan. It poses the question: is there any empirical evidence to suggest that the attacks are in some way a reaction to armed UMA strikes? Again the answer is far from conclusive, although there is some indication that the time window between armed UMA strikes and subsequent attacks on NATO convoys has been reducing.

It would seem that it is easy for people in Pakistan to say they are being radicalized by armed UMA strikes but little evidence emerges from factors that can be observed externally suggesting that is actually the case. For those tempted to suggest there is a measurable impact on the local people that drives them into terrorism, these results make difficult reading. As the armed UMA strikes are applied with increasing precision and focus more on people on the move in rural areas and avoid attacks in urban areas, the likelihood that people will be radicalized will arguably be reduced even further.

That, however, does not reduce the increasing political sensibilities on this subject. During a visit to Pakistan in August 2013 United States Secretary of State John Kerry seemed to announce an imminent end to UMA strikes in Pakistan. It appeared as though Kerry was playing to the gallery in Pakistan where the newly-elected Prime Minister Nawaz Sharif had demanded an end to UMA strikes. Within a matter of hours the State Department was forced into rushing out a clarification of the Secretary of State's remarks, suggesting that: 'In no way would we ever deprive ourselves of a tool to fight a threat if it arises.'

His view was presumably based upon the reports, some of which could be regarded as somewhat premature, that Al Qaeda's core leadership team in Pakistan has now been all but eliminated. Rationally that would seem to be an optimistic assessment. Al Qaeda's demise has been forecast too many times already. The movement, for that is what it is, shows a remarkable durability. While many presume that the current leader of Al Qaeda still resides in Pakistan, other key figures in the organization are also believed to be close by. These include Saif-ul-Ajal (the son of Osama bin Laden) and Azam Gadaahar, an American convert to Islam who has taken on an increasingly prominent role in developing Al Qaeda's narrative encouraging vulnerable people in the west to become lone wolves.

While these individuals remain at large it seems unlikely that Al Qaeda will suddenly reach a tipping-point and become irrelevant. That view has also been reinforced by the rapid geographical diversification of the movement using a model based on franchises. These now extend across a swathe from South-East Asia through the Middle East into North Africa and include insurgencies affiliated to the movement in places such as Tunisia, Libya, the Sinai Desert, Syria and Mali.

While the Al Qaeda franchises in Mali have been defeated by the French intervention in seizing the capital Bamako, they have dispersed and are simply waiting for the French forces to leave before attempting to take on the Malian army again. Next time, however, the insurgents will also have to contend with United States UMA strikes. Preparations in Niger to host armed United States UMA are already well advanced.

Cooperation between Israel and Egypt in the Sinai Desert using UMA to track Al Qaeda-affiliated groups operating in the area has also surfaced in media reports. Indeed, on one occasion in August 2013 suggestions even appeared that an armed Israeli UMA was used to kill four suspected militants in the Sinai. This is just the sort of remote area that Al Qaeda enjoys exploiting. They are adept at forging links with local Bedouin tribes and using those relationships to find shelter and sanctuary in areas where government control simply does not exist.

Historical insights on insurgencies

Today's insurgencies do not exist in an historical vacuum. Each draws upon the pantheon of past history and of examples where insurgencies have been successful and where others have failed. It is axiomatic that those that have succeeded have managed to tap into the emotions and support of local people. As many military commanders have correctly identified in contemporary warfare, the fight for the 'hearts and minds' of the local population is the centre of gravity of the campaign. Any events that give the insurgents the opportunity to manipulate the media simply have to be avoided if at all possible.

The issue for those engaged in the command chain of COIN campaigns is to do everything humanly possible to avoid civilian casualties, given what are inevitably difficult circumstances. Where things do go wrong it can hinder progress in the overall campaign. In what can be a feverish and emotional atmosphere in the wake of an attack where civilian casualties arise, the truth can often be submerged below a surface of rumours, lies and denunciations.

As opposed to their historical counterparts, today's insurgents enjoy many of the freedoms of social media. They have proved themselves to be adept at capitalizing on unfortunate civilian casualties. Securing the consent of the people can be difficult, especially when the people that may be killed in such operations often have close kinship ties with those holding important sway over the local people.

These events can fuel the insurgency and help maintain the tempo of hatred that is required to keep people fighting in what can often appear to be very difficult circumstances. Lessons from Iraq and Afghanistan show that COIN can be a war of attrition; it is just more selective regarding who ends up being killed. While comparisons with the kind of 'total war' of the early part of the twentieth century are difficult, military activities by all sides involved in a COIN campaign are often unable to distinguish the combatants from the non-combatants.

It may be a clichéd view but it is axiomatic that in war it is not just those engaged in combat operations who get hurt. For coalition commanders, in places like Iraq and Afghanistan the need to find other ways to maintain the pressure on an agile adversary was crucial. It was to Special Forces that the commanders in Iraq initially turned for help. They are specifically attuned to COIN.

The tactical evolution of the COIN doctrine that has achieved so much in Iraq and Afghanistan has built on the core concept of high-tempo operations by members of the Special Forces against key members of the command chain involved in directing the activities of the insurgency. This was where coalition commanders decided to place their tactical focus. As a tactic, it does carry risks. Non-combatants can be caught up in the exchanges, even when the intelligence sources being used are accurate and reliable.

UMA have also played their part in this process of dislocation, both at the strategic end of the equation in Pakistan – targeting the hideaways used by the Taliban commanders over the border – and in Afghanistan. By being able to operate in the difficult-to-access sanctuaries, the drones have played an important role in the psychological war that goes to the core of a COIN campaign.

Through exerting this kind of constant pressure against the second-tier commanders, the ability of the highest-echelon commanders to put their plans in place on the ground becomes disrupted and dislocated. The operations of the UMA have also clearly been shown to have a coercive effect upon the leadership of Al Qaeda, restricting their freedom of manoeuvre. The joint activities of the Special Forces and the Predator and Reaper UMA have had a major impact in helping shape the battlespace, creating the conditions in which some form of political reconciliation process can start. It is axiomatic in COIN campaigns that this rarely occurs quickly.

It would seem that only time, and a certain degree of war-weariness associated with the population at large, can create the conditions for an insurgency to start to lose the fundamental lifeline on which it depends: the people. Hence the average duration of insurgencies in history is around fourteen years, or half a generation. COIN operations rarely end quickly.

Kinship ties can also provide the insurgents with a high degree of immunization against any efforts by the counter-insurgents to create a narrative that exploits any fault lines that might appear in the course of routine insurgent activities. Even they make mistakes and provide the counter-insurgents with opportunities, albeit fleeting ones, to remind people that the insurgents are equally capable of hurting people. Indeed, the insurgents are far from innocent in this regard. The wilful use of children as young as 6 in the role of suicide bombers is an example of where they are prepared to act outside what might be regarded as socially acceptable.

Even when an insurgency is deeply socially embedded and reliant upon long-standing family ties, such exploitation of children under whatever theological pretext is available still creates a backlash. Sometimes the population involved in an insurgency must feel they are being pulled between two polarized forces, neither of which seems to respect the social norms, values and beliefs that provide the basic function of their society.

For the insurgents, maintaining the goodwill of the people is crucial. In this regard they have an initial advantage when it comes to securing the consent of the people, with their understanding of local customs, traditions and interpretations of customary law. Mao Tse Tung acknowledged this with his document 'Three Rules and the Eight Remarks' in which he laid out the nature of the relationship that insurgents should have with the people, placing it on a legal footing. His writing on this subject is the first time the idea of 'hearts and minds' was written down without it being formally referred to in that way. What Mao sought was a 'unity of spirit' between the insurgents and local people. He also benefited in trying to create that goal from a lack of the kind of cultural and religious schisms that punctuate contemporary COIN campaigns, adding a greater degree of complexity to mapping the societal landscape.

Mao specifically noted in the 'three rules' element of his doctrine that 'all actions should be subject to command'. The implication was that actions had to be approved and that no individual or lower-level commander could sanction actions that might create a schism between them and the local people. For the Taliban in Afghanistan the dislocation of the command chain has proven difficult for them to address. As the casualty count built up over 2009–2011, many second-tier commanders looked long and hard in the mirror. As their life expectancy decreased, so did their willingness to fight. For the COIN campaigns in Iraq and Afghanistan history will show that it was the relentless pressure placed on the command tier that finally created the conditions for some political process to begin.

Mao also addressed a wider set of points to lay out the way in which the population would be maintained on-side. Not stealing from the people or being selfish or unjust – the two other rules – provide complementary guidance on the nature of how to maintain the support of a population. The eight remarks provide a further amplification of the rules, noting that the insurgents should 'be courteous' and 'honest in your [sic] transactions'.

Without that support they risk, over time, becoming marginalized, isolated and eventually irrelevant, needing to resort to actions that try to coerce and intimidate the public into supporting their actions. For them this can undermine the complex social relationships that often modulate the way an insurgency works as apparent disagreements and long-standing blood feuds can be temporarily put aside in order to focus upon the common enemy: the counter-insurgent force. For those trying to build a complex societal map of the nature of the relationships this can be a tough problem as the links can change very dynamically.

For socially-embedded insurgents a reliance on kinship and family ties is important. They can provide the mask that helps screen their activities. However, over long periods of time even those relationships can wear thin as indiscriminate violence appears to bring no end to the conflict. Human beings, no matter what their culture and history of resisting invaders, can tire of war. The question for contemporary military commanders is how to sense that potential end game and how to maximize their chances of creating the conditions in which the wider public turns against the insurgents. This is the much-vaunted tipping-point for which military commanders charged with delivering some kind of politically acceptable outcome to COIN campaigns must now search.

This idea of creating a tipping-point harks back to the over-quoted supposed success story that was the 1950s British campaign in Malaya, dealing with the so-called Malayan Emergency. That a reasonable political outcome was achieved in Malaya cannot be in doubt. However, the methods used, creating what today would be referred to as internment camps, do not easily transfer in space and time. The world has moved on since then.

The geography of Malaya and the lack of media coverage of the relocation programme that saw thousands moved from their villages into camps were time-specific. This was how British High Commissioner Gerald Templer created the necessary dislocation between the population and the insurgency. He placed a physical distance between the two which was crucial. With the ubiquity of today's media, such an operation proposed by a leading military commander would not even pass first base in terms of political scrutiny. Creating tipping-points in a globalized and highly-connected world where respect for human rights is paramount has to be done without mass movements of populations. Effectively imprisoning large parts of a population is simply not a viable option in the court of international opinion.

Templer also understood, however, that creating a physical separation of the population and the insurgents was only part of the solution; he also recognized the need to create a mental separation. He essentially needed to divorce the population from the insurgents. His approach to this led to the development of what even today is still referred to as a 'hearts and minds campaign'. While in Malaya this twin-track approach to separating the population from the insurgents did succeed, the application of the same approach to contemporary counter-insurgency campaigns has not been so successful.

As the COIN campaign developed in Iraq in 2004 and 2005 the United States military turned to the lessons drawn from past COIN campaigns, such as Malaya, Oman and Kenya, for guidance and insights to inform their new doctrine of how to prosecute COIN campaigns. This initially saw a re-emergence of the use of the term 'hearts and minds campaign'.

Superficially the approach seemed one that might deliver a positive outcome. Time, however, was to prove its inadequacy. The societal landscape in Iraq was so very different from that in Malaya, Oman and Kenya. Iraq's tribal, clan and kinship ties were far more socially significant. Tipping the population in Iraq was going to prove troublesome. It was not just going to arise from using instruments that could be developed by the coalition military forces; some internal support from within the population would need to occur. Ironically the pressure for that to erupt came from the actions of the insurgents themselves, who created the conditions for a hugely reluctant population to seek assistance from the American forces.

Mobilizing the population against the insurgents, however, can be difficult as was shown by the Cyprus insurgency fought by the British between 1956 and 1959. This had a far less conclusive outcome because the British failed to prise away the population from the insurgents. This can be attributed to a number of things. Firstly the insurgent campaign in Cyprus was led by a brilliant and single-minded individual: Colonel George (Georgios) Grivas. Born in Nicosia in July 1898, Grivas was a local man who understood the importance of the societal and geographic landscapes of Cyprus and how that might create advantages for an insurgency. In Cuba, Fidel Castro applied similar ideas using the terrain of the country to his advantage.

Colonel Grivas's planning for the insurgency campaign was meticulous and took place over a period of several years. His strategy was to deeply embed the insurgency into the population, forging a bond that the British would find hard to break. His insurgency model is one that has some parallels with the way the Taliban currently operate in Afghanistan. The seven mountain groups were the people who bore the brunt of the fighting in the Troodos Mountains in Cyprus.

Alongside them he also created forty-seven town and seventy-five village groups providing the capacity to conduct small-scale reprisal attacks against people seen as collaborators or elements of the British forces, with assassination being their main tactic. Given those close social ties forged by Grivas, the Cyprus campaign never reached a social tipping-point. The commitment of the Greek Cypriots to the notion of enosis (political union with Greece) did not waver. In this situation the British found it very difficult to deal with the insurgency. Eventually, after a great deal of bloodshed, a political solution did emerge. A similar pathway was also found in Northern Ireland. In Spain, a political solution to the problems posed by ETA also looks increasingly likely.

Modern insurgencies

The problems with security along the border between Pakistan and Afghanistan have been well trailed in the media. For the Pakistani army the sheer scale of the security problems they are trying to deal with at the moment dwarfs the resources that can be allocated to security. Inside Pakistan several separate and complex insurgencies are at work. As a state, Pakistan treads a fine line between remaining viable and failing completely. The current situation there can be likened to a swimmer who is barely able to tread water in a rough sea. The situation in Yemen and Somalia is not much different. The French intervention in Mali was motivated by a similar situation. Here the government's authority was being challenged by another of Al Qaeda's highly-mobile franchises.

As Al Qaeda spreads out from its traditional bases in Pakistan, Yemen and Somalia, so American drones are being based at new locations to provide the means whereby they can be tracked and their activities disrupted. In early 2013 the creation of a new base in Niamey, the capital of Niger, is just one indication of how the trend towards wider geographic operations is developing. Developments in West and North Africa in places like Western Sahara, Mauritania, Libya, Tunisia, Algeria and Nigeria are all a cause for concern. Instability and a lack of governance in these regions is a magnet for those seeking to become involved in Al Qaeda's mission.

These areas also offer potential terrorists a large amount of manoeuvre room with several of the sanctuaries now forming part of a geographically-interlinked area. Previously Al Qaeda had to deal with its geographic centres of activity being dispersed. Now they can move seamlessly through a number of vulnerable countries. Mounting Special Forces operations in these areas to disrupt the activities of terrorist groups is therefore not straightforward.

After the successful French military intervention in Mali in early 2013 many of those involved in the insurgency have dispersed throughout the region. Reports have emerged of the remnants of those involved in Mali dispersing into Libya, Niger, Algeria and Tunisia. As Egypt seemingly descends into political turmoil, Al Qaeda-affiliated groups wait in the Sinai

Desert to strike at their adversaries. It may not be long before Israeli airplanes and UMA are flying over areas of the Sinai that they gave up in the hope of achieving a long-term peace deal with the Egyptians. What goes around seems to come around.

This can only have a detrimental effect upon the overall security picture in the wider Middle East and North Africa. Those dedicated individuals displaced by the French intervention in Mali are unlikely to just give up their vision of creating Islamic states ruled by their interpretation of Sharia law. Egyptians who think the west has somehow operated in concert with their military forces are likely not to draw a line at street protests. For many, Al Qaeda's long-term vision is still a dream that can come true. It is all a question of creating the right conditions.

For the international community the problem posed by states such as Mali is a difficult one. Whereas the French chose to intervene militarily for a short period of time to put down a threat to the very existence of the Malian government, such as it was, other countries are increasingly reluctant to place soldiers on the ground. Vietnam produced its own long-term legacy for the American military. That is now being repeated after the long-term wars in Iraq and Afghanistan. The situation in Syria provides a warning to anyone who thought that military operations in Libya were somehow a model that could readily be repeated.

In Mali the French government with support from its European partners is trying to develop a model of military intervention that closely resembles the operations mounted by British forces in Sierra Leone in 2000 when Operation PALLISER was authorized. In Mali the French enjoyed the benefits of close cultural ties to enable them to quickly establish a footprint on the ground. Support from NATO colleagues also helped. The Royal Air Force flew over fifty heavy-lifting missions using its fleet of C-17 aircraft to move forces into Bamako when the initial decision to deploy was taken by the French government. A similar speed of response occurred in Sierra Leone when an insurgency threatened to bring down the internationally-recognized government.

The insertion of ground forces from the Afghanistan side of the border into Pakistan is difficult. The operation to kill Bin Laden shows the scale of what is required if Special Forces are to try to neutralize the threat from the Taliban and Al Qaeda groups operating in the border regions of the NWFP and the FATA. Only a small number of passes are navigable and these are well-known to the insurgents who use them as supply lines into Afghanistan and also to evacuate people wounded in the conflict in Afghanistan for

131

treatment on the Pakistani side of the border. Putting ground forces into these areas for any length of time is hugely risky. The teams involved could be compromised on both insertion and extraction, if both of these phases are to be done by land.

Helicopter insertion or extraction is hazardous, as the mission that resulted in the death of Bin Laden showed. One of the helicopters involved crashed as it was unable to sustain the necessary lift due to a variance in operating temperature at the point of attack from what was predicted. The insertion or extraction by helicopter is complicated further by the altitude at which any insertion would have to be made. At such altitudes the thinness of the air creates huge problems for lift. Noise also carries for long distances, alerting the insurgents that a military activity is under way.

The attack on Al Qaeda elements hiding in the Tora Bora area of Afghanistan on the border with Pakistan highlighted these issues. Operation ANACONDA was an example of the kind of issues that any helicopter-based mission into Pakistan would face for insertion, extraction and any ongoing re-supply missions. Re-supply by air, using low-level flights to air-drop munitions, water and food also had attendant risks. If any of the air-drops were not completely precise, their activities would compromise the mission.

While during the Second World War the Allies maintained a deep mission inside Burma, the fact that troops were operating behind enemy lines was obvious from the military operations in which they were engaged. The Japanese knew the forces were operating in the jungle; it was just a question of finding them. The bare mountains of the Hindu Kush offer no such cover.

The tribal landscape inside the FATA is also unlikely to be friendly or provide shelter to a covert force trying to operate inside Pakistan. While it is clear that some people in the villages in the FATA have little regard for insurgents and for their international guests from far-off places like Chechnya, they also would be very wary of being drawn into the middle of the ongoing dispute between the warring parties. For many people living in the FATA they just wish all of the protagonists that have decided to come and live in their area would go away so they can just get on with their lives.

The only way Special Forces could operate inside Pakistan in the FATA is if they operated in really small teams and were able to live off the land without any support from across the border. If they were wounded or compromised there would be little chance that help would be on hand in the kind of time needed for an extraction if they had come into contact; a military euphemism for being engaged by an enemy.

This discussion illustrates the dilemma behind the increasing use of armed

UMA for attacks on potential terrorist strongholds, no matter where they are established. If large-scale military intervention is not politically or economically acceptable, then other methods must be found to remove the kind of threats that led to 9/11.

The situation is actually very simple. If such people are allowed room to plan large-scale mass casualty attacks, they will carry them out. While events in London on 7 July 2005 were small by comparison with the scale of casualties in America in 2001, they arose because people were able to visit terrorist training camps in Pakistan. Simply saying that border controls should be enhanced to monitor people leaving Western Europe for destinations such as Pakistan, Somalia (via Kenya), Yemen or Syria is not enough. Somehow the training camps must be held at risk. As has been shown, the options to do that at the moment are somewhat limited.

The results of armed UMA attacks in Pakistan are rarely far from the headlines. The use by the United States of their advanced Predator and Reaper UMA to attack targets in the remote and inaccessible parts of the FATA was on an upward spiral. The much-reduced use of armed UMA attacks in Pakistan following the peak in 2010 is down to a combination of political pressure and a general recognition that many of the senior commanders in the Taliban have already been removed. Signature attacks – where behavioural indicators are used to decide whether activity on the ground has terrorist overtones – have also been scaled back. These are the most likely to cause the kind of collateral casualties that reignite emotional responses inside Pakistan and have political repercussions.

While the armed UMA attacks do have their notable success stories, such as when the Taliban leader Baitullah Mehsud was killed on 5 August 2009, more often than not the outcome of the attacks creates a backlash; in this case people in various parts of Pakistan taking to the streets to complain about their countrymen being killed. The sense of grievance on the streets of Karachi and Islamabad is palpable. Fed by rumours that often quickly gain traction in such societies, the people's anger is genuine and also reflective of a wider range of concerns they have about their security.

The attacks increase tensions in what is already a fractious relationship between the United States and Pakistan. It is an uneasy alliance that was further complicated by the obvious lack of trust that existed between both parties at the time of the raid by United States Special Forces that resulted in the death of Osama bin Laden. This took a situation where disagreements were largely voiced and managed in private to a new level when the Pakistani government, feeling humiliated that its ally had acted on its soil with no forewarning of an attack, decided to act; expelling members of the United States intelligence services and ground support teams supporting the operations of the Pakistani army.

The relationship is one fed by the necessity to cooperate in defeating a protean threat but wracked by arguments over how to accomplish the desired outcome. Diplomatic cables released by Wikileaks provided important insights into the underlying nature of the relationship between Pakistan and the United States, with political leaders' public and private views being thrown into stark relief.

Both parties agree that the lack of security in the FATA is a source of major problems. However, that is about the extent of the agreement. Trying to develop a consensus on how to solve the problem is a very different matter. The United States keeps urging Pakistan to do more to disrupt and destroy the insurgent and transnational terrorist sanctuaries in the FATA. Pakistan counters those diplomatic and military messages with its own rhetoric of the manpower it has devoted to trying to create security in an area where the inhospitable terrain plays into the hands of the insurgents.

The Americans justify the attacks as being part of a strategy that intervenes upstream when a clear and present danger exists to their security. If the people being targeted are not put under some form of military pressure they will be free to plot new massive attacks against the United States homeland. The spectre of a nuclear detonation hanging over New York is a vision that haunts even the most passive members of the United States administration.

Their argument is that they simply cannot take the risk of allowing new terrorist training camps to develop. The transnational terrorist groups that use these facilities have developed powerful arguments concerning the theological justification for the use of weapons of mass destruction against the United States. Taking any risk that one day these people could carry through on that threat is simply not possible. Indeed, it would be totally irresponsible for any government to ignore such a potentially catastrophic outcome. Where such a threat exists, it has to be neutralized.

The problem is that transnational terrorist groups gravitate to ungoverned spaces. The FATA, famed for its independent history, is just such a place that transnational groups like to inhabit. The mountainous terrain of the area poses huge difficulties for anyone trying to conduct ground-based military activities. It provides an ideal sanctuary for the insurgents fighting in Afghanistan and for groups operating there who have a broader international agenda.

Therefore as far as Washington is concerned these ungoverned spaces cannot be allowed to go unmolested. The images of 11 September 2001 are simply too raw in people's minds to allow another event to be planned from within the FATA. The American public would not be very forgiving if another attack, aimed at raising the bar from the almost 3,000 who died in New York, Washington and a remote field in Pennsylvania, was to occur.

The question is how to tackle the threat that exists in the FATA. There would appear to be four options for addressing the problem of the insurgents and their international guests. They apply in Pakistan and in other countries where UMA are starting to be used to attack people and groups affiliated with Al Qaeda.

These are as follows:

1. They are removed from the area by the Pakistani government mounting a major military operation into the specific areas of the FATA where the groups have taken refuge. This has been tried before in South Waziristan with an initial level of success.
2. They are removed from the area by internal militias (*Arbakai*) formed by the local people to defend their territory from external threats. This is complicated by the nature of the relationship between the Pakistani government and the local militias, which has important historical overtones.
3. They are engaged and destroyed by coalition ground forces projected into Pakistan either overtly or covertly to target specific training camps operating inside the FATA in places such as North Waziristan. This includes the use of cross-border artillery salvos into villages thought to be hosting insurgents and terrorists.
4. They are engaged and destroyed by the use of air power, one option being to employ armed UMA.

Each of these four options has its advantages and disadvantages. These vary at the strategic, operational and tactical level, both militarily and politically. In the course of the book we shall analyze these in some depth to document the issues and derive a balanced view of their relative merits from the perspectives of the parties involved. An analysis of these varying issues highlights the reasons why the United States has increased its focus on drone operations in Pakistan and elsewhere, such as in the Yemen and Somalia.

From an American viewpoint, to allow the training camps to fester is to invite another attack. The do-nothing option simply carries too many risks.

Of all the various diplomatic and military options available, the use of drones seems the most cost-effective in an age where austerity is a word rarely far from the lips of political leaders in the western world. The same argument also applies in Israel when it comes to the increasing presence of Al Qaeda in the Sinai Desert and Syria.

From the viewpoint of the Pakistani government and other governments that also come into the cross wires of a burgeoning drones campaign, the potential for civilian casualties and a backlash against the political leaders who appear to condone such activities is also a high-risk venture.

Of course this could be partially solved if the nature of the targeting by the drones could be so precise that only insurgents or their immediate families would be killed, eliminating the deaths of innocent civilians altogether. That is a utopian dream. It is axiomatic in war that sometimes civilians will die, caught in the crossfire of implacable opponents.

The issue is whether the level of civilian casualties can be kept below the threshold of pain that sees an entire society rise up against the attacks. In Pakistan while there are a vocal few, that the media always picks up on, there are other more pragmatic groups and individuals who recognize that solving the problem is not going to be easy and the pain of civilian casualties has to be accepted.

The reality of UMA operations
On 27 September 2010 members of the International Security Assistance Force (ISAF) crossed the border from Afghanistan into Pakistan in what is known as 'hot pursuit' of a number of Taliban insurgents who were trying to escape to the sanctuary of North Waziristan, having attacked an ISAF outpost in Khost. Supported by Apache gunships with their Hellfire missile systems, the ISAF is reported to have killed forty-nine insurgents. Reporting available at the time suggested that no civilian casualties occurred.

The geography of the local terrain in the border region between Khost and Pakistan would seem to back up the claim that no civilians were killed in that particular attack. The insurgents would have been caught out in the open and the outcome of the engagement would have been conclusive. Asymmetric warfare is easy when the insurgents have somewhere to hide. If they cannot simply disappear, the odds swing back decisively in favour of the side with the greatest weapon systems. Small-arms fire, unless it gets lucky, does not deter a determined Apache helicopter pilot.

This incident is one of a small number that have occurred in the immediate vicinity of the border between Afghanistan and Pakistan. Once insurgents are able to travel 15 miles inside North Waziristan they can find some shelter in the small villages that exist along the many riverbanks in the area. Settlements, with varying numbers of compounds, provide a sanctuary from which they can re-arm and launch themselves again across the border. Targeting these villages with the aim of disrupting the insurgents' activities and also killing the senior leadership of Al Qaeda who are thought to reside in the area has become the task of the Predator and Reaper drones operated by the United States military.

Equipped with Hellfire missile systems, these small objects have been increasingly used in remote areas of Pakistan and places such as the Yemen to target people who are linked to transnational terrorist groups and their franchises. Operating around the clock, these airborne robot warriors are being increasingly used to attack the sanctuaries in which the insurgents involved in activities in Afghanistan and Pakistan reside. In September 2010 in the run-up to the ninth anniversary of the terrorist attacks in the United States a notable 'uptick' occurred in the frequency of attacks mounted in the FATA of north-west Pakistan. In remote areas the leadership of the Taliban and Al Qaeda were being hunted down and killed.

Civilian casualties

While aimed at a wider range of states using such methods, the comment from the United Nations report obviously chimes with other concerns that have existed over the use of drones in Pakistan and other countries. At the heart of these concerns is the discussion of how many innocent civilians are killed by drone attacks. The lack of transparency on this matter does not help and is an area where the United States may be forced to give ground in the court of world opinion.

Estimates vary wildly as to how many people have died in the attacks. The insurgents are often quick to react and cordon off the areas preventing anyone, even local people, from gaining accurate information on the numbers of people actually killed and injured. With little opportunity for any form of external validation of the casualties it is difficult to decide on the relative merits of the competing narratives emerging from the United States and the contrasting views offered by the insurgents.

Occasionally the attacks are reported to result in large numbers of civilian casualties, creating a new momentum behind the debates on their use. In a case reported in May 2010 by Al Jazeera and other networks, senior military officials were disciplined after a drone attack in February 2010 reportedly killed twenty-three innocent Afghan civilians. The report said that four

American officers had been disciplined by General McChrystal for 'inaccurate and unprofessional' reporting. Since that event the United States has stepped up its efforts to reduce the numbers of civilian casualties that can be directly associated with the use of drones. Senior officials are clearly worried about a media backlash against their use.

Attitudes to the drone strikes in the remote valleys and regions of Pakistan are also difficult to fathom, with local factors making opinion polling unreliable. Gallup, conducting a survey for Al Jazeera in July 2009, showed only 9 per cent of the 2,500 Pakistanis interviewed across the country supporting the use of drones. More recent polling carried out by the Regional Institute for Policy Research and Training reported in the Swat Valley on 16 May 2010 showed that 67 per cent of the 384 people asked felt that the drone attacks 'provoked' the families of those killed. Revealing research conducted on the ground in the FATA region of Pakistan published in September 2010 by the New America Foundation suggested that 70 per cent of the people interviewed wanted the Pakistani army to tackle the problems of the terrorists, with support for drone operations only being voiced by 22 per cent of the population.

While the overall picture emerging from the opinion polling appears to show a clear rejection of the use of drones and great concern for the reported levels of civilian casualties, the truth on the ground may, in practice, be more complicated. There are several rival viewpoints.

Noor Behram is a native of North Waziristan. In July 2011 an exhibition opened in London of pictures he had taken in the immediate aftermath of drone attacks in Pakistan. These provide graphic evidence of the problems that arise when civilian casualties are caught in the crossfire. They are almost bound to evoke an emotional response.

Farhat Taj, a native from the north-west region of Pakistan working in Norway, provides a contrasting viewpoint in her regular commentaries for *The International News.* She comments that: 'Most of the literature misinforms in terms of civilian casualties caused by the attacks.' She reserves her most stinging criticism for the New American Foundation who have published assessments in the early part of 2010 indicating that 32 per cent of those killed are innocent civilians.

Farhat Taj offers a contrary viewpoint based upon discreet conversations with people living in the area whose views differ markedly from the widely-held perceptions published in the media. Through her own contacts and on the ground visits she has established that there are many people on the ground in North Waziristan in favour of the use of drones. They see them as a way of ridding themselves of the strictures, privations and intimidation associated with the violent acts of the insurgents, many of whom have killed local tribal leaders who have dared to oppose their actions.

While it is hugely difficult to estimate the numbers of people killed by the Predators and Reapers and compare that with the local people in North Waziristan killed by the insurgents, reports emanating from the region suggest that the insurgents have been conducting operations against local people; many of whom have been described as spies, feeding information to the Americans operating the drones.

In a situation reminiscent of the conditions existing before the spontaneous uprising of the Awakening Councils in Al Anbar province in Iraq, the local people's patience in North Waziristan may be being severely tested. In Al Anbar it was the increasing use of violence against local people that eventually created the conditions for the Awakening Councils to eject the insurgents from the local area.

By taking its time the Pakistani government may be hoping for a similar reaction from the local people in North Waziristan whose tribal loyalties are based upon somewhat similar customs, traditions and creeds. The devil, however, is in the detail and many in the tribal systems in North Waziristan feel their allegiance to their customary tribal code of Pashtunwali, which is deeply ingrained in their values and beliefs systems, makes it extremely difficult for them to take similar steps to the tribal uprising in Iraq.

In Pakistan the Lashkars (local village militias) have a similar role to that undertaken by the tribes in Al Anbar. When local people need to be protected it is the local elders and tribal leaders who come together to decide on raising a Lashkar to provide security for their area. However, their success to date has been patchy, forcing the United States to maintain their drone operations.

Farhat Taj's arguments are based on a view that Al Qaeda and Pakistani Taliban-aligned groups operating in the area have sought to systematically kill many of these leaders to ensure that Lashkars will not be raised against them, creating a dislocated and fragmented response at local level. This is a lesson that Al Qaeda learned from events in Iraq. Many of those involved in creating the 'awakening' have since paid with their lives as the remnants of the insurgency in Iraq have sought revenge in a series of targeted assassinations aimed at the leaders and their families. By turning quickly to the tactics of targeted assassinations, the Pakistani Taliban and the remaining elements of Al Qaeda operating in the FATA in Pakistan have disrupted the deployment of Lashkars.

Therefore it is unlikely in the short term that the kind of systematic and coordinated tribal uprising that so fundamentally shifted the balance of power in Al Anbar in Iraq away from the insurgents is likely to occur in Pakistan. What are euphemistically referred to as 'shaping operations' are therefore likely to be an enduring requirement until the situation changes significantly, with the Predator and Reaper drones at the forefront of those ongoing operations.

The targets attacked by the armed UMA can be broadly divided into two groups. The first comprises fixed locations such as dwellings and the second is made up of mobile targets. These occur against either urban or rural backdrops. For the operators of the armed UMA, the fixed urban location is the most difficult.

People can be seen entering a building but understanding just how many people are there and what they are doing is difficult unless reliable and robust human intelligence sources exist. This is where the targeting approach is most at risk. Simply deciding to attack an urban dwelling because a number of armed people have walked in to what appears to be a meeting is not, in itself, a sensible decision. People meet together in places like Afghanistan and Pakistan for a number of quite legitimate reasons. These meetings can also be of an ad hoc nature, i.e. there is no obvious time element to when they occur.

Attacks on rural dwellings also have their problems. Gaining accurate human intelligence in such situations may risk compromising the source. Patterns of behaviour, however, can be built up over a period of time. The presence, for example, of a specific vehicle at the location may be strongly indicative of a specific high-value target being there. Attacking remote rural locations does, of course, reduce the risk of civilian casualties.

However, looking at the kind of structures that exist in these areas, what may happen is that a family creates an initial dwelling and then as their numbers grow, new members' accommodation is built onto the side of the existing compound. In rural areas of North Waziristan, for example, the pattern of land ownership is often reflected in the ways in which compounds are laid out. Large gatherings of people in such a compound are not directly indicative of terrorist activity. It may simply be a family event to celebrate a birthday or wedding. Discriminating such innocent events from those with evil intent is really difficult unless specific indicators, such as the presence of a particular car and/or individual, increase the certainty over what is happening inside the compound.

Increased attention to all of these factors has had a clear impact on the targeting strategy that is emerging. An increasing number of the armed UMA strikes that are now reported in the media mention attacks on cars or motorcycles. These are the terrorists' preferred method of transport. In Afghanistan the Taliban often use motorcycles to move quickly between locations. Targets can often emerge quite quickly and can be fleeting in their appearance, placing pressure on the decision-making apparatus. In the Yemen and Pakistan a detectable shift in the targeting strategy is emerging with a preference for engaging targets on the move in rural situations, minimizing the risk to civilians in the area. With a car on the move the operators controlling the armed UMA have more time to choose the point of engagement. Of course, targets on the move do pose problems, even for the high technology behind armed UMA strikes. The missiles can miss. Human intelligence may also provide information that suggests a certain individual was seen getting into the car but once the vehicle moves it is up to the UMA operator to stay with the potential target until the strike is authorized and the location for the attack is deemed suitable.

This analysis of the evolving approach to targeting, however, is all based on a simple assumption: that the targeting of such individuals and groups, be they foreign citizens or American, British, German or any one of a large number of other nationalities, is legal.

Despite reassuring noises emanating from the White House, the idea that the use of armed UMA strikes is not legal as far as international law is concerned is an issue that does not seem ready to disappear. The United Nations has issued a report that was critical of the use of UMA, questioning their legality as a weapon of war, even when used in successful attacks against specific insurgent leaders such as Baitullah Mehsud and Mustafa Abu al-Yazid on 21 May 2010. The report notes that the drone attacks have 'had the very problematic effect of blurring and expanding the boundaries of the applicable legal frameworks.' Particular attention was drawn in the United Nations report to the issues of 'human rights law, the laws of war and the law applicable to the use of inter-state force'.

The debate that will inevitably result from the publication of the United Nations report is unlikely, in the short term, to have an impact upon the rate at which drone attacks are being carried out. However, for an American president so clearly aware of his image and that of his country, the arguments of legitimacy may tell over a period of time. The main observation arising from the United Nations was that the failure of states to 'provide transparency and accountability for targeted killings is a matter of deep concern'. Clearly in early 2013 these issues finally started to bring a response from the White House as the veil of secrecy over armed UMA strikes started to be lifted.

The utility of the Predator UMA
The image of the Predator drones has changed dramatically in recent years. From being a long-duration ISTAR platform focused on detecting targets that other forces could prosecute, drones are now hailed in some parts of the media as semi-autonomous robots that are capable of finding, fixing and destroying targets of opportunity as they arise. An air of invincibility has grown up around the Predators. They can go into areas where the deployment of Special Forces could prove politically difficult. Their inherent flexibility has brought their capabilities to the attention of many other countries who are interested in acquiring such systems. The reality of the situation, however, is somewhat different.

The Predators are not the kind of devices that Hollywood sometimes likes to depict, in reality being quite vulnerable. In places such as Pakistan the Predators operate against the backdrop of a rugged and mountainous terrain where the local people are almost constantly aware of their presence. The engine noise, barely discernible at sea level, is audible in the remote hillsides of North Waziristan. Local people call them wasps, with reports suggesting that up to five may be on station at the same time.

The Taliban and their associates hiding in North Waziristan have managed to deploy some anti-aircraft weapons and have claimed that they have shot down a number of the Predators operating in the area. Technical faults have also had an impact upon the programme with reports suggesting loss rates of up to 40 per cent of the Predators, with specific difficulties emerging during take-off and landing. It would seem these are not the super-weapons some in the media would have us believe.

Nevertheless, this form of upstream activity is becoming ever more prevalent as links between terrorist groups in Pakistan and potential attacks in the United States, such as the failed car-bombing in Times Square, have been established. The ten-count indictment returned in the Southern District of New York charged the man suspected of the bombing, Faisal Shahzad, with 'conspiring with the Pakistani Taliban to wreak death and destruction in Times Square'. Transnational terrorist groups continue to evolve their tactics and with the attempt to bring down an airliner on Christmas Day 2009 over Detroit being planned and developed in the Yemen, it is not difficult to imagine drones being deployed against terrorist training camps known to be operating in places like Marib Province in the Yemen.

UMA have become the weapon of choice when it comes to maintaining pressure on terrorist groups operating in their remote hideaways in places such as North Waziristan and in various parts of the Yemen. The inherent flexibility and agility provided by the UMA makes them an ideal solution in contrast to the political fall-out that may occur if more combat forces were to be committed to countering the increasingly geographically dispersed operations of Al Qaeda. They are therefore unlikely to stop being used in the short term. Indeed, it is possible to argue that when General Petraeus sanctioned Special Forces operations in up to seventy-five countries where terrorists are currently thought to be operating, it is very likely that UMA will become more widely used.

With Al Qaeda franchises continuing to provide security challenges across Africa and South-East Asia and the emerging links between its franchises in the Maghreb and groups such as Nigeria's Boko Haram, a broader base of operations being conducted by drones seems likely. Given their utility in disrupting the activities of terrorists, who try to use the complex geographic and societal landscapes to take refuge, it would be difficult to suggest that drone technologies will suddenly become regarded by military and political leaders as having passed their sell-by date. If anything, by their definition of utility, their time has come. The question is: in a highly-connected world where images of people killed in drone strikes have the capability to radicalize people previously unwilling to become involved in terrorism, is the cost of the utility too high?

Creating tipping-points for a population

For military commanders seeking to understand how to create a tipping-point for a population, they might note the words of William Shakespeare when he attributed the following quote to Brutus in the play *Julius Caesar* (Act 4, Scene 3): 'There is a tide in the affairs of men which, taken at the flood, leads on to fortune; omitted, all the voyage of their life is bound in shallows and in miseries.'

For military commanders the ability to sense that tide in the affairs of war when a population is reaching breaking-point requires their antenna to be tuned to some very different forms of indicators to those that usually define progress in a military campaign. While in the First Gulf War one metric, that of the number of Iraqis deserting the military and returning home, was a clear indicator of the morale and imminent collapse of the Iraqi army as a coherent force, measuring a similar rate of conversion of the population away from supporting the insurgents is far more difficult.

Despite a great deal of scholarly analysis and doctrinal development (much of which has occurred in contact), COIN operations remain challenging. The classic technical solutions that seemed to provide the answer for military forces engaged in large-scale warfare, such as the encounters documented by many commentators of the various battles that led to the swift end of the First Gulf War in 1991 and the relatively quick end to the Second Gulf War, no longer seem appropriate. The potential of Intelligence, Surveillance, Target Acquisition and Reconnaissance (ISTAR) assets to provide what many commentators saw as a golden age of decision superiority has remained largely unfulfilled. The potential for the all-seeing eye to provide great situational awareness to commanders who can then exploit that position to outmanoeuvre an adversary has been outwitted on too many occasions.

In the immediate aftermath of the Second Gulf War the nature of warfare took a new turn. Within Iraq many who had been disenfranchised from their traditional position of power decided that it was necessary to oppose the occupiers. Slowly but surely, an insurgency developed that eventually enveloped Iraq. Large-scale military activities, such as brigade sweeps, failed to suppress the activities of the insurgents who rapidly gained in strength as some of the actions of the military forces trying to secure a new and stable Iraq in the wake of deposing Saddam Hussein alienated local people, creating a distance between many of the population and the occupying forces.

The insurgents quickly learned to manipulate the media. Extreme displays of violence against captives and hostages were broadcast on the internet. The insurgents also developed their tactics, avoiding direct clashes with the occupying military forces and developing a wide range of Improvised Explosive Devices [IEDs] to create havoc among both the civilian population and their military guardians.

In using such tactics, attempts were made to open the fault lines that had the potential to fragment the Iraqi population as sectarian violence was meted out in indiscriminate attacks across the country. The inevitable backlashes that the attacks were designed to provoke did occur and the violence threatened to descend into anarchy. Pictures beamed back into the sitting rooms of populations in the west gave rise to a feeling that the situation was hopeless; Iraq had moved to being an ungovernable state. The groundswell of opinion in favour of leaving Iraq grew and created a political momentum for action.

Recently, as a result of a range of actions by the United States military, their coalition allies and many brave people in Iraq, that situation improved for a short period of time. The country still suffers from acts of extreme forms of violence but the drumbeat is lower and the trends are positive. The turning-point in the campaign came in 2006 as the United States learned that it had to change its approach to COIN operations. While that provided the platform for some new ideas to be applied, the 'cat was out of the bag' as far as asymmetric warfare was concerned. COIN had moved into a new form designed to exploit the leverage offered by the media, targeting a global audience, many of whom had hoped at the end of the Cold War that warfare was now a thing of the past.

The period in Iraq from 2004 to 2007 and the ongoing situation in Afghanistan were to conclusively prove the false premise on which those hopes were based. Colonel Thomas Hammes is one of a number of authors who have provided in-depth insights into this emergent form of warfare. His book *The Sling and the Stone: On War in the 21st Century* uses a clever analogy of David and Goliath to show how insurgencies have changed; developing the idea that contemporary COIN operations are a fourth generation of warfare. He chronicles in detail the changes from the earliest forms of insurgent warfare developed by Mao Tse Tung to the complex situations that arose in Iraq, although the publication of his book in 2004 means that many of the insights he offers are necessarily tempered by the history emerging from that point.

His analysis of Afghanistan, describing it as a tribal network, was important. His observation that 'the United States actions proved that a nation can still surprise 4GW insurgents' is important; reflecting as it does on the way in which the early part of the Afghanistan campaign relied on Special Forces and their intelligence agency counterparts to support the indigenous Northern Alliance to mobilize and eventually defeat the Taliban. His question at the end of the chapter that considers Afghanistan is, however, prescient. He asks: 'Does the coalition, particularly the United States, have the political will to sustain a decades-long effort?' While history is still being written in 2013, it would appear that the answer is no.

The political reality of the world at the start of the second decade of the twenty-first century is one where military operations are unlikely to be conducted with the same level of enthusiasm as they were in the first decade, when some political leaders appeared to believe they were on a moral crusade to bring democracy to the world and deliver an internationally stable situation. The inspiring vision of a 'universal civilisation' described by the Nobel Laureate V.S. Naipaul appears distant. The works of Samuel Huntingdon, notably the picture he painted in *Clash of Civilisations*, are more relevant to the contemporary world.

What had previously been confined to the kind of local and somewhat remote conflict in Sri Lanka, where the Tamil Tigers developed a variety of forms of asymmetric warfare, was now available to anyone across the world who wished to access specific material on the internet. For COIN, which had previously been seen in a quite different light in campaigns in Malaya, Oman, Kenya and Northern Ireland, a new paradigm of warfare was about to appear: one in which the notion of decision superiority and of being inside an adversary's decision loop was to prove less of an advantage than developing a very specific understanding of the local customs, creeds and traditions that modulate the local societal landscape.

The centre of gravity of COIN operations

Throughout the history of COIN, one enduring truth remains: the local population matters. They are, to use the military term, the centre of gravity of the campaign. It is they that can help create the conditions in which the insurgents can become marginalized and alienated from the people. It is the civilians who can, in effect, say to the insurgents: 'You no longer have a role in this society; you are irrelevant.' This was the set of circumstances that provided the game-changer in Iraq as the leaders of the tribal councils in Al Anbar Province rejected the approach to governance offered by Al Qaeda based on their strict interpretations of Sharia law. There are many people in Afghanistan who have similar reservations regarding these interpretations. In the aftermath of the attacks on the United States on 11 September 2001 and the military operations in Afghanistan, it is fair to say that many in the local population did not lament their passing when they were initially defeated.

While progress has been made in Iraq, its history has yet to be written. The insurgency shows an enduring ability to maintain a toll on Iraqi citizens. While its support and infrastructure is greatly diminished and disrupted and the Iraqi security forces are showing great progress in running and conducting their own affairs, there remains a stubborn underbelly of those disenfranchised from power who refuse to accept the political processes that now govern Iraq. There may yet be a great deal of history to write about the Iraqi insurgency, much of which will be bloody and costly in lives and economic progress. Another fact about COIN: it rarely comes to an obvious end-point at which anyone can stand up and claim to be the victor.

In a highly-connected world where media frenzies await the unwary, avoidable civilian casualties have their drawbacks. Albert Schweitzer, the renowned German philosopher and theologian, perhaps got closest to this thought when he remarked: 'Revenge ... is like a rolling stone, which [sic], when a man hath forced up a hill, will return upon him with a greater violence, and break those bones whose sinews gave it motion.' Schweitzer's words, said when warfare was a very different proposition, seem to chime ever more readily in today's complex world of ethnically and religiously motivated warfare.

Where societal customs demand retribution when family members are killed, unintended consequences await those who act in haste: there have been too many cases where as a result of such actions people have had time to repent at their leisure. Pakistan is one of a number of countries whose societal landscape can be said to be complex. To ensure the actions taken in Pakistan are carried out in ways that minimize the potential for adverse and unintended reactions, it is vital that anyone contemplating conducting any form of military activity in its remote and tribal areas understands the nature of this dynamic landscape.

Since the early days of COIN operations the consent of the civilian population for the activities of the military has been crucial. In what is arguably one of its first inceptions in China, Mao Tse Tung insisted on 'a unity of spirit' between the local civilian population and his military forces. To provide guidance on his intent in this matter Mao Tse Tung wrote his 'Three Rules and the Eight Remarks' document which essentially defined the rules by which his military forces should operate on the ground.

John Nagl, the author of the excellent analysis of COIN operations *Learning to Eat Soup with a Knife*, observes that the 'implementation of such precepts allowed the army of the people to be truly an army of the people.' Mao Tse Tung understood the importance of having good relations with the local population if progress in COIN activities was to be made. His use of the analogy of 'fish swimming in the sea' also vividly portrays the notion that the local population provides the environment through which the insurgents can swim and operate.

Western military forces have struggled to understand the dynamics of the societal landscapes involved in Iraq and Afghanistan. Initially it was assumed by some that the experience gained by the British army in places like Malaya, Oman, Kenya and Northern Ireland in fighting colonial-based wars would provide a huge repository of knowledge from which to draw. The term 'hearts and minds campaign' is one often cited in the media. While a simplistic notion, and one with far too many historical connotations, it conveys a sense of the aims of the military: to ensure that the local population provides support for the activities designed to create a secure environment for social and economic development and give some leeway to the military forces when things go wrong. War is, after all, unpredictable.

While it remains embedded in the language of COIN activities, it is possible to argue that the use of the term 'hearts and minds campaign' is no longer appropriate in the twenty-first century. The language has moved on, with the focus now being on gaining and maintaining the consent of the local population for the military activities. This is less ambitious language. It is a pragmatic recognition that the language of 'hearts and minds' more befits a time when the media was absent, such as in Malaya, and the military could build entire new settlements in which to house people and create a physical separation of the local people from the communist insurgents.

In Iraq walls did have to be built, temporarily (and in Northern Ireland), to separate people whose intent of sectarian violence was all too clear to see. Al Qaeda's franchise in Iraq to this day undertakes operations designed solely to promote sectarian recriminations. So while populations were not physically displaced by the military as in Malaya, they were separated; although many chose to leave their homes in mixed Sunni-Shia communities in places like Baghdad because of the intimidation brought about by gangs of militia with specific sectarian allegiances.

Unfortunately the one definite lesson to emerge from the past is that it appears that each COIN operation is subtly different from the last. While superficially many similarities existed between Belfast and Basra – a journey down any of the streets of Basra bringing back specific memories of similar reactions from similar patrols in Belfast – the reality underneath was so very different. It was too easy to think, from a superficial viewpoint at the outset of the campaign, that the sectarian divide between Catholic and Protestant in Northern Ireland had an immediate and obvious parallel in the religious divisions between Sunni and Shia populations in Basra and Baghdad. Unfortunately this was far from the truth.

To understand COIN you need to look below the veneer of what appeared to be a sectarian conflict with remarkable similarities to Belfast. In Basra it was not a bipolar world that, once mastered, was relatively easy to understand. It was a protean landscape where alliances and affiliations between groups could change in an instant. Therefore this was a very different form of COIN operation.

These differences, however, while difficult to detect, have a huge implication for the development of the COIN doctrine. It did not take too long for the United States military to go back to the drawing board and start re-writing their entire approach to asymmetric warfare. The blood and treasure invested by the United States military in Iraq demanded nothing less than a complete overhaul of the ideas. Field Manual 3-24 was the result of that effort. It remains the product of a huge intellectual effort by highly-qualified and experienced people who set out to capture and document the experiences of Iraq and lay a platform for future COIN operations.

Given the nature of asymmetric warfare and the sheer freedom of manoeuvre now enjoyed by our adversaries, it is unlikely not to be updated in the near future. Time moves on, as do the tactics and approaches to asymmetric warfare. We now live in a highly-connected world. What worked (albeit temporarily) for the Tamil Tigers in Sri Lanka may well, with a few subtle modifications, work anywhere across the world. Asymmetric warfare, where our adversaries target our weakest points, is unlikely to be a thing of the past any time soon.

While the doctrine of courageous restraint is being advocated on the ground by the military in Afghanistan, the apparent indiscriminate nature of the attacks mounted by drones creates a whole new perspective on warfare. Courageous restraint places a huge burden on the war-fighters on the ground. They are being encouraged not to be too quick on the draw: to hold back on squeezing the trigger, even when provoked, thereby putting themselves in greater danger in order to reduce the numbers of civilian casualties. This point is crucial if the consent of the local population is to be secured and maintained.

COIN operations do not win the hearts and minds of the local population if civilians appear to be killed at random. This is where armed UMA strikes are vulnerable to the accusations levelled against them. The single most challenging point for the ISAF leadership in Afghanistan is the issue of civilian casualties. It is the open wound into which each further civilian death pours salt, creating a hugely emotional climate in which progress towards security becomes more difficult, if not in some cases impossible. It is something that Afghan political leaders return to on a frequent basis.

This is why the coalition military in Afghanistan are asking their military forces to put themselves even more in harm's way. However, there is a huge discontinuity at the heart of this strategy. The CIA with respect to their operations in Pakistan does not seem to be singing from the same hymn sheet as the military in Afghanistan. The complex thread of tribal ties that runs through the social fabric from Pakistan into Afghanistan increases the danger of retribution reaching across geographic borders. It also highlights the huge fault lines that exist in the approach adopted by the United States with respect to the way it is conducting its attacks using drones. There seems to be one rule for the military on the ground and another for the intelligence agencies and the drones.

DRONE WARFARE

It is easy to understand the emotional reactions to the deaths of seven members of the Central Intelligence Agency (CIA) in Khost on 30 December 2009; always a particularly difficult time of year, being in the Christmas/New Year period. Those who carried out the attack knew the trigger it would provide for retribution. For the CIA it was a very bad day. They lost several of their most experienced officers in the attack; people who were hugely difficult to replace. The reaction of the CIA in stepping up armed UMA strikes, however, would give momentum to a cycle of renewed violence, the end of which is hard to see. The recent cycle of strikes initiated in the Yemen may have similar consequences. This problem of cyclical violence is something that Alfred Schweitzer would have understood.

Virtual wars

War used to be up close and personal. Now it is remote and disconnected from the reality on the ground. Accusations that those prosecuting the armed UMA strikes are detached from reality and think they are involved in playing out a war game have proven difficult to refute. Warriors no longer look directly into the eyes of their opponents. The classic Western shoot-out at dawn so stereotyped by Hollywood no longer applies. The marshal and the bandit do not test each other to see who is fastest to the draw. Or do they?

They work through packages of sensor systems that convey images halfway around the world to screens manned by people who have just driven to work, having had breakfast with the children. Technology is creating a new class of warrior, bringing a whole new aura to the idea of a beltway bandit. However, the technology is only an enabler of war. Until robots do develop largely autonomous capabilities to conduct warfare, a situation also predictably massaged by Hollywood in a number of films, man will remain in the loop.

Despite being thousands of miles away, the man still has to make the decisions; to call the point at which the predator will strike. Yet unlike the wolf who stalks his prey, often in a pack, waiting for the prey to become weakened and knowing instinctively when to strike, the modern-day technological predator relies on the instincts of the person flying the drone. His or her instincts come into play as they seek to track down high-value targets in remote and inaccessible parts of the world and attack the locations in which they hide. Their predatory instincts are reminiscent of the wolf and yet are quite different, being driven by human emotions and reactions. Their opportunity to kill is in part motivated by revenge. For the wolf, survival is the key preoccupation.

The crucial point for the human predator is the point at which to launch the weapon; in the Western movies the point at which the gunslinger goes for his weapon. The speed of the draw is all-important. It defines the status of the person. It is axiomatic that he who draws fastest survives, as long as he can shoot straight. In many stereotypes that little point is often forgotten as the drama is played out.

Taking this analogy of the shoot-out so often central to the storyline in a Western film, the operator of the drone does have to think about when to go for his weapon. His advantage is that the target cannot immediately return fire. This suggests that the modern-day gunslinger operating the drones should be patient, taking the model from the animal kingdom and waiting to strike when the enemy is at his weakest. Unfortunately this is where the analogy breaks down. The military gunslinger cannot wait for the right moment.

In many cases fleeting opportunities to target specific high-value assets such as leaders of the insurgency have to be taken at very short notice. With people becoming increasingly aware of their need for operational security, such as avoiding the use of mobile phones, the opportunities to attack and kill certain key leaders in the insurgent movement can quickly pass. Anecdotal reporting has suggested that on several occasions Osama bin Laden, the alleged leader of Al Qaeda, had escaped while hierarchical and cumbersome command and control systems debated the legitimacy of his being targeted.

The use of the Predator drones to conduct these precision attacks, guided by what is called 'actionable intelligence', is not, however, pain-free. There are consequences. Drones do not provide the other intelligence collection opportunities afforded by using Special Forces. The latter are able to retrieve important additional sources of material – so-called 'pocket litter' – from sites where operations have been conducted.

The flexibility afforded by the drones, allowing remote locations to be attacked, is important and clearly resonates with President Obama. Where multi-national terrorist groups hide, the drones can search them out and destroy them. They can avoid the need for troops to be placed on the ground; helping deny organizations like Al Qaeda their aims of drawing the United States into a wider set of bloody conflicts with potentially huge economic costs. For this reason, and several others, it is likely they will remain the 'only game in town' for some time to come when it comes to trying to disrupt upstream activities by terrorist groups.

CHAPTER 7

Into the Future

The scene on the deck of the USS *George Bush* looked quite normal. The date was 11 July 2013. Crews were in their positions awaiting the arrival of the next aircraft in the circuit to land. From the media coverage the whine of the engine on finals could be heard. Suddenly there was a brief glimpse of the aircraft shooting past the television cameras as it successfully picked up the second wire laid across the deck. The television camera then panned away along the deck to show the aircraft that had just landed.

This was not the usual arrival on the deck of a United States aircraft carrier. With little fuss or bother the X-47B prototype from the next generation of fighter jet taxied away off the landing area. It had cost $1.8 billion and eight years of development work to get to this point. Yet what was amazing about the whole scene was the apparent normality of the event. It was as if it had happened hundreds of times before.

In reality that was not the case. It was the first time an unmanned aircraft had ever completed that feat. The world of UMA had just moved into an entirely new era. The world of armed, unmanned fighter jets operating from aircraft carriers had just arrived. Airmen watching the event may have wondered how long it would be before they no longer experienced the rush of the catapult. As if to drive home the point about the impending end of manned platforms, the UMA repeated the same feat minutes later as it landed for a second time on the deck of the aircraft carrier. It was only on the third approach that the UMA developed a fault that required the landing to be aborted. The UMA then flew away to land safely at a nearby shore-based facility.

For the next generation of naval airmen the images of *Top Gun* would no longer capture the vivid nature of air-to-air combat. The landing of the X-47B presaged a new future in naval aviation, one that perhaps did not have quite the same level of excitement that their forefathers had experienced. From this point on the United States navy had entered the era of UMA. It was a significant moment. UMA were starting to venture beyond the land environment into the maritime domain.

Maritime applications

UMA are now not just being applied in the land environment. In December 2013 a Los Angeles-class attack submarine launched a UMA from one of its vertical missile tubes. The launch from the USS *Providence*, which was the first Los Angeles-class submarine to be equipped with vertical launch tubes for anti-shipping missiles, was a success. The UMA was encapsulated in a Sea Robin launch vehicle which separated as the package surfaced. The UMA then deployed its wings for flight and conducted a two-hour surveillance mission broadcasting real-time video back to the submarine which remained submerged for the exercise. The launch was the result of a six-year development activity led by the Naval Research Laboratory.

The test vehicle used on the exercise has the potential to fly for up to six hours helping develop the recognized land, littoral or maritime picture in support of reconnaissance or combat operations. It has an obvious role supporting the activities of Special Forces. Such a capability does, however, have its limitations as the launch or presence of a UMA might reveal that the launch platform has to be nearby and may result in the submarine's position being compromised.

In the marine environment UMA are also involved in monitoring criminal behaviour. One variant that started its life in the land environment is the ScanEagle. In 2012 it completed 600,000 combat hours. Of that total, 23,000 hours were spent operating in the maritime environment on around 3,000 sorties. It is not the only UMA that has been adapted for use in the marine environment. The MQ-8B Fire Scout is a helicopter-based UMA that surpassed 5,000 flying hours in April 2012.

In one of its first major deployments at sea the MQ-8B was based on board the USS *Halyburton* and the USS *Simpson*. Plans have been announced to arm the Fire Scout with a laser-guided 70mm rocket. The next generation of the system moves away from the smaller platform using a modified Bell Model 407 helicopter. The first flight test model of this was delivered to the Naval Air Station at Point Mugu in California on 8 July 2013 and was slated to make its first operational flight in the autumn of 2013. In service this will offer increased payload capabilities (40 per cent), range (30 per cent) and endurance (100 per cent). The MQ-8C retains 85 per cent of the flight control software used by the MQ-8B.

The initial contract with the United States navy sees fourteen of the MQ-

8C being supplied alongside the equipment for seven ground stations. On board the new UMA a sensor system called the Multi-Mode Sensor Seeker (MMSS) provides the ability to look for targets in the maritime environment, such as small pirate skiffs or mother vessels such as dhows. The programme is also taking the first steps towards increasing the degree of processing on board the UMA, reducing the need for streaming video. A database on the UMA coupled with automatic target recognition software will enable some pre-screening of the data. The system, it is claimed, will be able to look for specific ships.

This sudden interest in the application of UMA to the maritime domain had one important driver. Off the coast of Somalia as the problems with piracy grew rapidly the international community turned to UMA to provide the kind of persistent response with which they had excelled in Iraq and Afghanistan. What was needed was to patrol large areas of the Indian Ocean looking for indications of the presence of PAGs (Pirate Action Groups) and also to provide ISTAR support over specific events. The imagery derived from a UMA operating over the small dinghy in which Captain Richard Phillips was held hostage for several days by armed pirates came from one that had been deployed in support of the operation. It provided vital situational information that enabled the rescue of Captain Phillips to be successfully accomplished. The platform in question was the ScanEagle system. This had originally been designed to help fishermen locate and track schools of tuna.

Since its initial development the ScanEagle UMA has been trialled by the Canadian navy and the Royal Navy. Aboard HMCS *Charlottetown* in the Mediterranean Sea the UMA played an important role in helping gain situational awareness data as part of Operation ACTIVE ENDEAVOUR, the mission to bring security to the people of Benghazi in Libya. The Royal Navy has also conducted trials of the same UMA on board HMS *Sutherland* and deployed the same system aboard a Royal Fleet Auxiliary during exercises in the eastern Mediterranean Sea. In June 2013 the Royal Navy announced a major contract with the manufacturers of ScanEagle to deploy the system at sea.

The Royal Navy categorizes this UMA as a Maritime Unmanned Air System (MUAS). The ScanEagle system can travel at speeds of up to 80 knots (92 miles per hour, 150 kilometres per hour) and can communicate with its host platform up to a range of 100 kilometres (62 miles). One test variant of the platform has achieved a record of remaining airborne for twenty-two hours and eight minutes. When it returns on board it is captured by a 'Skyhook' retrieval system.

The ScanEagle system can be configured with a number of different sensor systems to suit specific missions. It is also envisaged that it might be possible to extend the range over which such UMA operate by providing a control console in indigenous aviation assets, such as Merlin helicopters, that are deployed on board destroyers and frigates. They could also receive the direct read-out from the UMA sensor package.

The problems of monitoring large areas of the Indian Ocean dwarf the uses of UMA in land-based theatres. This is a different level of surveillance altogether. The sensor suite aboard the UMA over the ocean was not optimized for a maritime environment. Radar reflections behave differently over a developed sea to how they do over land. This requires several of the existing sensor suites deployed on UMA to be re-optimized for the maritime environment.

Basing UMA in the Seychelles was an obvious solution. Their geographic location was ideal for flying surveillance missions over those areas of the Indian Ocean where PAGs were known to be active. This deployment lacked the intensity of media coverage associated with armed UMA in places like Pakistan, the Yemen and Somalia.

The Predators operating out of the Seychelles provided another point of pressure against the pirates, restricting their operations. However, it is unlikely that anyone is going to be writing any features suggesting that the deployment of UMA against the pirates had anything but a marginal impact on their operations.

What did change the entire dynamic in the region was the introduction of armed guards on merchant vessels and the deployment of secure rooms into which the crew could retire when they came under attack. As long as the crew could hold out for up to a day, naval vessels could reach the hijacked merchant vessel and effect a rescue. In the limit armed UMA could be used to halt a pirate attack but what is more likely is that the lightly-armed tactical UMA could be used to threaten a PAG if they continue an attack. The vision of a pirate surrendering to an armed drone may not exist in the imagination for much longer.

The role performed by UMA over the Indian Ocean has reaffirmed the role they can play in maritime security operations. Australia is a country that has a huge coastline to protect and has problems with illegal immigration. Japan has issues with China over the ownership of the Diaoyu Islands. Mexico has a drug-smuggling problem. All are actively in the process of acquiring UMA capabilities to patrol vast areas of the ocean.

The BAMS (Broad Area Maritime Surveillance) system is one new development that will increasingly allow UMA to play a role in policing international maritime boundaries. African nations, such as Nigeria, will no doubt soon be following suit. For European countries the problems of criminal groups smuggling economic migrants, potential terrorists and narcotics from the shores of North Africa to the southern shores of Europe is a growing concern.

In terms of naval strike capability the United States navy has already started the UCLASS programme, awarding a number of the main US defence suppliers initial contracts to develop designs for the programme. Using a UMA to project power into the littoral or over the horizon from an aircraft carrier is not a great leap of faith, although the amount of ordnance that can be carried is limited compared to the F-18 Hornet.

For other missions, such as anti-submarine warfare, the UMA will have to be armed with different weapon systems. With the MQ-9 Reaper already carrying 500lb bombs as part of its payload, the weight of a contemporary torpedo such as the Stingray (267 kilos, 500lb) suggests that arming a UMA for an anti-submarine strike mission is not out of the question.

The size of the Stingray torpedo would present a design challenge for the teams involved in developing an ASW (Anti-Submarine Warfare) capability but it is unlikely the issues that arise would be insurmountable. Replacing the F-18 Hornet in the air-to-air combat role, however, is likely to be a significantly greater challenge. The anti-ship strike role is also not an ideal environment for Hellfire, the small warhead being more appropriate in a COIN context. Against a warship it lacks the capability of missiles like the Exocet with its 165 kilos (364lb), although the speeds of the two missiles are not at great variance. The Hellfire travels at 425 metres per second and the Exocet at 315 metres per second. From a kinetic energy viewpoint, which scales at the square of the speed, both have a significant capability to punch a hole in a ship.

In August 2013 it emerged that plans were being developed to equip UMA with their own air-to-air weapons for defensive purposes. The aim is to equip the MQ-9 Predator and presumably the Reaper system with the AIM-9X Sidewinder, AIM-120 Advanced Medium-Range Air-to-Air Missile (AMRAAM) and the High-Speed Anti-Radiation Missile (HARM). Alongside the missiles the MQ-9 would also be equipped with an Active Electronically Scanned Array (AESA) radar derived from those used on the most advanced fighter jets in the United States Air Force inventory.

The primary aim of the initial studies that are examining the feasibility of this configuration is to add a counter-UMA mission to the work already undertaken by the aircraft. The addition of HARM would also give the UMA an ability to conduct the SEAD mission. It is also possible to see in the future the UMA acting as the scout or pathfinder for incoming packages of strike aircraft relaying targeting coordinates directly into the cockpit of aircraft such as the F-22 Raptor. This would provide the advantage that the F-22 would be able to remain stealthy and not illuminate the target using its own on-board radar system and risk being compromised.

China is also showing interest in the development of maritime UMA. Images emerging on the internet showed a Chinese frigate – the Jiangkai II (Type 054A) vessel *Zhoushan* – launching a rotary-wing UMA. Its design resembles that of the Camcopter S-100 developed in Austria. The company manufacturing the S-100 denies selling the S-100 to China. It is possible that the close resemblance of the two is entirely coincidental or it may be another example of how the Chinese have taken steps to accelerate their own development programmes using espionage to obtain designs and drawings from which they have been able to quickly engineer their own models.

The S-100 is capable of carrying a payload of up to 50 kilos (110lb) and can remain airborne for up to seven hours. In April 2012 it became the first UAV to fly from an Italian warship: the Artiglieri (Soldati)-class frigate ITS *Bersagliere*. During the flight tests the S-100 operated in sea states varying between 3 and 4 and at wind speeds up to 25 knots. In contrast to the MQ-8C, the company manufacturing the system in Austria (Schiebel) have made it clear they do not intend to arm the S-100. Its role is purely as an ISTAR asset, although it can also carry loudspeakers, spotlights and rope/net-dropping containers to try to have an effect upon a target, such as a pirate skiff.

The uncertain security landscape
As more and more UMA are developed the uncertain international security landscape provides increasing opportunities for their use. It was in June 2010 that President Obama secretly authorized a massive extension in the use of search-and-destroy missions by United States Special Forces. After he had taken office *The Times* reported that these forces had been operating in up to sixty countries.

The decision taken by the president increased that list to seventy-five states. It came as it emerged that United States Special Forces had killed thirty-four out of the top forty-two Al Qaeda commanders in Iraq in a surge of activity aimed at disrupting their operations. Similar results were seen in the Yemen when six of fifteen Al Qaeda commanders were also killed. In parallel with these operations President Obama had also authorized an increase in the level of armed UMA strikes in Pakistan. This was a determined effort to bring Al Qaeda to its knees.

The organization, however, is not taking the onslaught from UMA strikes lightly. It is trying to fight back. Material discovered by French military units in Mali in early 2013 provided what was tantamount to a manual informing potential insurgents how to avoid UMA strikes. Osama bin Laden had also been known to correspond with Al Qaeda franchises on the subject.

In Al Qaeda chat rooms requests for 'brothers' to come forward with ideas on how to defeat the threat from UMA strikes have been published. The first issue of the magazine *Azan* published by the Taliban in Afghanistan and Pakistan also appealed for help in countering the threat from UMA. The American Defense Intelligence Agency is even on record as suggesting that Al Qaeda is conducting research into developing jammers for satellite navigation signals derived from the GPS constellation. In 2012 researchers at the University of Texas showed just how easy it was to 'spoof' the navigation system on a UMA. In Iran one of its most senior commanders has even gone so far as to suggest that schoolchildren are being trained to spot the signatures of UMA and provide warnings of their presence over sensitive nuclear-related facilities in the country.

In an ever-changing world, prescience is a valuable albeit risky attribute. Reputations are easily lost by those who make confident predictions about the future, only to be proved wrong. As in the past with UMA, what had to be balanced is what can be forecast from reasonable extrapolations in technological development and the pull-through that emerges from operational requirements. In the short term that is not too difficult. In the medium to longer term, complications arise.

This is not to suggest that the development of UMA somehow grants a licence to print money. After years of work and a large investment, Germany suddenly announced in May 2013 that it was not going ahead with its planned purchase of RQ-4 Global Hawk platforms that had been intended to replace its aging fleet of Atlantique ATL1 twin-turboprop SIGINT aircraft. It had intended to buy five platforms and operate them from Schleswig-Jagel Air Base on the Baltic coast.

The RQ-4B Block 20 variant of the Global Hawk had been the foundation of the programme. It was to have been equipped with a European-developed SIGINT package. Ground stations built in Europe would have received the direct read-out from the platform as it circled over an area of interest. One of the reasons cited for the cancellation was a decision reached in America to phase out the Block 30 variants of the Global Hawk.

In the United States Congress has been careful to scrutinize the various programmes being undertaken by the Pentagon. The Block 30 RQ-4 Global Hawk platforms were slated for retirement as Congress questioned the cost of their operations in comparison with the manned U-2 reconnaissance aircraft. This is despite their having achieved a number of important operational milestones such as the 30,000 combat flying hours in 1,500 sorties in February 2010.

The manned platform was seen to be able to operate at a higher ceiling, out of the way of the weather systems that can upset the flights of the RQ-4. The ISTAR suite on the RQ-4 was also reported to be a slightly inferior derivative of that flown on the U-2. With an original fleet of sixty-three aircraft having been planned, the first change in the programme saw that scaled back to forty-five.

By the end of September 2012 fourteen were in service with four more in production. Each aircraft was supposed to cost a reported $35 million. However, reports emerging from the United States suggested this had risen very significantly to over $200 million per platform. In a time of austerity such increased costs were bound to trigger higher levels of scrutiny of the programme.

However, as their plans developed the United States Air Force shifted its planned procurement of the Block 40 variant (RQ-4B), moving the total being purchased to forty-five aircraft. The Global Hawk programme now appears to be something of a political football, being metaphorically kicked between Congress and the United States Air Force. Congress has told the USAF that it should keep the Block 30 variant in service until the end of 2016.

The history of UMA has shown that developments in its core enabling technologies have eventually led to a wider range of missions and applications of the platforms. Developments in radio and flight and navigation systems have all had an impact upon the operational versatility of UMA. From being a simple unguided missile with dubious accuracy around the time of the First World War, UMA established themselves in the role of targets for gunnery practice. That required specific developments in flight control systems. To deliver unguided attacks, however, further developments in navigation systems were required.

Droning on
One mission that is likely to remain in the portfolio of UMA is that of being a target drone. To test the manoeuvrability of the next generation of air-to-air missiles, realistic scenarios will have to be created. The target aircraft will

have to be able to fly in ways that simulate the capabilities of current and emerging fighter jets. The QF-4 is one such example. It is based on the McDonnell Douglas F-4 Phantom II combat aircraft. A total of 300 have been converted since the start of the programme in 1996. Although the main tests involving the QF-4 have involved air-to-air testing, some have also participated in air-to-surface missile developments.

In order to keep up with the developments of the next generation of fighters, such as the Sukhoi PAK-FA and the Chinese J-20, the QF-4 target drone is being replaced by the QF-16. A contract to convert 126 of the former F-16s into target drones has been placed by the Pentagon. This will provide the next generation of targets for aircraft such as the F-22 Raptor which is the first fifth-generation fighter in service in the world.

As the fifth-generation fighter jets move from development into production it is vital for air forces around the world to evaluate how their aircraft would perform when up against the next generation of the threat. To meet that need the next generation of aerial target is already being introduced into service.

Similar capabilities are offered by the Mirach 100/X transonic aerial target system. It can stay airborne for over 100 minutes and can fly at a maximum speed of just over Mach 0.92. It is the latest generation of aerial targets that have been under development in Europe since the early 1980s. It is launched from a trolley system, making it highly portable. Two JATO boosters provide the power plant. The Mirach 100/X is able to fly at altitudes of 3 metres up to 12,500 metres. This enables it to provide a range of realistic threat profiles covering missions conducted by strike aircraft, fighter jets, sea-skimming and cruise missiles and UMA. It also carries a variety of payloads that can simulate stealth coatings on fighters and strike-bombers. Real-time telemetry is also transmitted to a ground station.

Today with the power of GPS navigation systems have arguably reached a plateau in their development. Guidance systems and other flight control dynamics are also areas where development work is likely to be restricted on current airframes used in the land domain. Such is the state of development of these areas that two RQ-4 Global Hawks can now be flown in such close proximity as to be capable of in-flight refuelling operations.

The KQ-X activity will shortly demonstrate in-flight refuelling between UMA. That will be a significant enabler for increasing persistence still further. This is a remarkable development and one that shows the maturity of UMA flight control systems. Flying in close formation is not easy as turbulence generated by the lead aircraft can have an effect on the second platform. Interestingly, in the case of the KQ-X trials it is the second aircraft that acts as the source of the fuel. This is quite different to normal air-to-air tanking operations.

Unmanned strike capability

Operations from the decks of aircraft carriers will, of course, add in new dimensions to the requirements for flight control systems. Landing on a moving deck in a variety of weather conditions has its challenges, both for manned and unmanned platforms. Now that the X-47B has completed several such landings successfully, the United States navy can begin to evolve its own ideas as to what the next generation of its combat aircraft should look like. The requirements for that are being developed under the UCLASS programme.

The United States Air Force equivalent programme MQ-X had stalled with senior officials suggesting they would wait and see what emerged from UCLASS before deciding what would happen next. Their caution is important. Developing an unmanned ground-attack capability to operate in non-permissive environments is possible. Speeds are increasing and the stealth characteristics are important from a survivability viewpoint. However, creating a replacement for the F-22 air-to-air fighter is a very different proposition.

Perhaps with the development of missiles that can pull G-forces beyond the physiological capabilities of the human body, the era of the air-to-air manned dogfight is coming to a close. UMA would act as a host platform for the missiles, moving them to the edge of contested airspace waiting for tasking against enemy forces by command and control systems. Any plans to replace the F-22 with another manned aircraft are bound to raise concerns about the potential costs involved. Similar drivers will also affect the development of future strategic bombers.

Talk of the next generation of long-range bombers being unmanned, however, is likely to prove premature. The Long Range Strike-Bomber (LRS-B) is seen as the replacement for the B-2 Spirit, the B-1 and the venerable B-52 bombers that are currently in service with the United States Air Force. The prohibitive costs of the B-2 saw the operational total set at twenty-one. Somewhat fortuitously this coincided with the end of the Cold War.

Had a third world war ever broken out the B-2 would have been tasked with two important missions. One would have been to attack major Soviet command and control nodes. The other would have been to track down land-based mobile missile systems. In the early days of any campaign the generation rate of the B-2 force would have been stretched. As things turned out the B-52s and the B-1s were to play a slightly different role from that for which they had been originally built. Their contributions to the First and Second Gulf Wars and the initial attacks in Afghanistan supporting the Northern Alliance forces against the Taliban often involved bombing tactical targets using a variety of conventional weapons.

Over its in-service life the B-2 also contributed to operations in Iraq, Afghanistan and in Libya. These have only required a small contribution from the stealth bomber in its geo-strategic role. Its ability to fly extremely long ranges and return home to the United States after delivering its payload have shown the potential for this kind of intercontinental strike-bomber. However, fearful of a repeat of what happened regarding the cost of the B-2, the Pentagon has imposed a cost-per-aircraft cap of $550 million. Trade-off studies to determine the advantages and disadvantages of manning the LRS-B or operating it as a UMA are no doubt already under way. In considering the design of the LRS-B the really difficult part is to try to envisage how the international security landscape will evolve over the next fifty years.

Versatility will be a key word in shaping the design. With America now paying far more attention to the Pacific Rim, there are scenarios that envisage conflict with China. In that kind of eventuality the LRS-B will have to be capable of operating in a variety of permissive and non-permissive environments. The issue of operating from a stand-off position or having to penetrate enemy air defence systems is going to be an important design consideration.

As the speeds at which UMA fly also start to increase as their use in air-to-air combat starts to be considered, some evolution of the airframe is inevitable. The Pentagon MQ-X programme covers the design of a UMA that can operate in contested airspace. As the designers look carefully at the MQ-X a number of factors will be important. Stealth will remain a key characteristic as intelligence collection operations have to take place in increasingly hostile environments. Both of the designs that were evaluated for the United States navy demonstration programme on UMA had stealth-like features. The RQ-180 UMA is the latest embodiment of these design characteristics and towards the end of 2013 was reported to be in testing at the Groom Lake Air Base in Nevada, home to the infamous Area 51 where the United States Air Force tested the U-2 spy planes in the 1950s.

The design of the RQ-180, depicted on the cover of *Aviation Week*, shows an aircraft that has a close resemblance to the X-47B. It is reported to have a range of 1,200 nautical miles and be capable of flying for up to twenty-four hours. Its primary mission is intelligence, surveillance and reconnaissance but it could be adapted to conduct electronic attack missions and carry other equipment to help suppress the effectiveness of an adversary's air defence systems.

The 'bat-wing' shape and the lack of any rear stabilizer are obviously linked to the configuration of the X-47 and contribute to the stealth characteristics that are so prized by designers of contemporary systems. With no ability to protect themselves against attack by surface-to-air missiles or interdiction by air-defence fighters, their only hope of surviving in a non-permissive environment is to remain undetected. However, as air defence systems continue to improve to counter stealth there will come a point at which UMA will have to be capable of self-defence. That might come in a number of ways, including using a defence pod that carries countermeasures designed to defeat homing air-to-air or surface-to-air missiles.

In the United Kingdom the preliminary designs of the Taranis UMA, named after the Celtic god of thunder, also closely resemble the features of a stealth aircraft. Indeed, the outward similarity between the Taranis and the RQ-170 Sentinel UMA is striking. UMA seem to be following a distinct design pathway that involves the creation of stealthy platforms.

The Taranis is being marketed as an Unmanned Combat Air System (UCAS) demonstrator. If it were to form the basis of a next generation of fighter jets it would have to be able to operate both as an ISTAR asset and perform combat duties in non-permissive environments. The RQ-170, however, is clearly designed as an intelligence collection platform whose service ceiling is reported to be around 50,000 feet. Around twenty of the RQ-170s are believed to be in service, although one was lost on operations over Iran.

As far as other technical issues go the rate of development of radio systems, particularly those involving satellite communications, are unlikely to introduce dramatic new developments in the short term. Bandwidth will still be at a premium. Situations where UMA have to be able to contend with being out of touch with their home base will still arise. The decision-making programmed into the control systems on the platform have to factor in a number of operational considerations. If radio communications are briefly interrupted the platform can choose to enter a holding pattern for a set period in the hope that communication links will be restored. If the outage lasts longer, then some simple mathematical analysis needs to be explored to evaluate options. Fuel state, all-up weight and distance to the nearest operational airdrome all need to be considered.

The picture is different when it comes to developments on the mission payload and weapons front. Missile warheads will evolve from their current capabilities. Programmable warheads will allow further improvements in the delivery of precise effects. Sensor systems will also continue to improve as hyper-spectral imaging systems provide even greater degrees of target discrimination and resilience to the kind of simple countermeasures that can sometimes thwart the targeting process. Increased resolution of the targets will also help.

Where missions monitor pattern of life behaviour before engaging a target, this will also help reduce collateral damage. UMA may not, however, be the solution of choice on all occasions. In the United States the decision to maintain the manned U-2 reconnaissance platform over the Global Hawk is one that may have potential ramifications. While UMA have many advantages in being able to go into environments where sending men is undesireable, cost is still an issue.

Mission evolution, however, is not going to be an area where dramatic developments will take place in the short term. The core roles of UMA in suppressing enemy defence systems, providing a target drone for gunnery and air-to-air interception practice, collecting intelligence and providing ground attack will remain. Use of UMA for maritime strike operations is the one area where new developments are already taking place. The days when UMA are developed to fulfil the kind of manned airlift capabilities offered by the C-130 and C-17 aircraft of today are well into the future; not that it would be difficult to deliver such a capability if a pressing military need arose.

It is principally at the other end of the spectrum where dramatic developments are being made in UMA. Smaller, lighter, insect-like platforms are already being developed for applications at the tactical level of command. Being able to covertly monitor the activities of an adversary while beaming back images to a control station has many attractions.

This is particularly true where military commanders seek to reduce the potential of collateral damage to a minimum. Having the ability to look into a room and see who is present is a beguiling prospect. If such a capability had been available before the assault on the compound used as a sanctuary by Osama bin Laden it would have provided some valuable additional information for the Special Forces involved. The risk is, of course, that the attempt to collect intelligence up close to the potential target becomes compromised.

Developments in these areas are all under way and the outcomes and short- to medium-term trajectories are not too hard to estimate. Where things become complicated is in trying to estimate the degree to which UMA will be increasingly able to conduct autonomous operations. It is reasonable at this point to suggest that a clear demarcation will exist between ISTAR and defence suppression missions and those involving any form of strike upon a target.

This is where a major problem exists. The implications of reducing sensor-to-shooter cycle times to a minimum provides an obvious driver to increased autonomy of target selection and engagement on UMA in the future. That, however, requires imagining that it would be straightforward to somehow programme the rules of engagement into a mission computer on the UMA.

This may sound easy on paper but in practice is not so simple. Any targeting decision arises from a lengthy analysis of a range of issues. Legal supervision is crucial if the legitimacy of the operations is to be defended in the court of international opinion. In addition, considerations of the target and its surrounding areas are not trivial tasks that can be readily automated. Moving to a situation where targets are identified and engaged without some form of human overview is not a solution that is politically acceptable at present or one that is likely to emerge any time soon. That does not, however, prevent researchers delving into the problems involved. Just as in the early days of UMA developments, the issues have to be identified and potential solutions devised.

For defence suppression missions the picture is different. Writing software that is able to adjust to changing parameters being sensed in the operations of radar systems is not impossible. Already jamming systems have a huge degree of automatic capability. However, as radars become increasingly digitized, so their flexibility and ability to adapt to a complex EW (Electronic Warfare) environment will tax the rate at which jamming signals can react. What was once a relatively simple game of measure and countermeasure in the EW field is becoming increasingly complex.

The sheer variety of waveforms and operating characteristics of digital radar systems will dramatically change the ways in which EW is applied in the future. To some extent automation of the response is inevitable, given the dynamic that is unfolding. That, however, is not the sole problem for the UMA defence suppression mission. Things can get quite complicated in other areas.

Take, for example, the kind of mission that involves several UMA cooperating with each other to suppress a defence system. If one is lost to enemy action the others have to re-programme the scope of their operations to provide a continuing capability. With their power budgets likely to be limited, optimizing the approach is going to require either a fully distributed decision-making capability or reference to a central command node that will change mission parameters, having decided upon an optimum configuration to account for the combat losses.

This will also have to operate in a dynamic environment where strike packages accompanying the defence suppression platforms are successful in removing enemy radar systems. As the nature of the threat changes, so the escorting platforms will have to adjust their priorities. Conserving power is also likely to be an important factor.

Increasing automation

Increasing automation is one of the inevitable design drivers for UMA. The RQ-3 DarkStar platform was one of the first examples of increasing levels of automation. The designation of RQ shows that the platform was primarily designed for unmanned reconnaissance activities. It had an unusual asymmetric wing arrangement with a total wingspan of 21.3 metres (69 feet). The prototype of the RQ-3 made its first flight on 29 March 1996 and crashed shortly after take-off. Its design incorporated stealth technologies. This made it difficult to detect and able to operate in non-permissive environments.

Its key difference, however, was that it was fully autonomous. Given the fears over UMA being able to act on their own and be capable of killing people, the use of this kind of language to describe its capabilities is in fact an exaggeration. It was able to take off, fly a mission profile incorporating several targets to be observed, operate the sensor payload, transmit real-time imagery to a ground station and recover to a pre-determined location on the ground.

In practice this so-called autonomous operation is quite limited. It is important not to assume that the RQ-3 was able in some way to think for itself. What it was able to do was follow a carefully worked out mission profile. Facilities also existed for a ground operator to update the mission profile during a flight should some revision of the mission plan be required at short notice.

The range of the RQ-3 has been reported to be just below 1,000 kilometres (621 miles) and it was able to cruise at a relatively slow speed of 464 kilometres per hour (288 mph). It had an operational ceiling of 13,500 metres (45,000 feet) and relied on its stealth technology to avoid being engaged when flying over an adversary's territory.

This places it somewhere between a tactical and operational asset as far as military commanders are concerned. Its development programme was, however, abruptly cancelled in 1999. The reasons for this are not clear but it is likely that the RQ-3 was not seen to offer anything specific when compared to some of the other UMA that were being developed at the time. The first prototype is on display at the National Museum of the United States Air Force at Wright-Patterson Air Force Base in Dayton, Ohio. Another prototype is on display at the Smithsonian National Air & Space Museum in Washington, D.C.

Looking to the future, the idea of a fully autonomous UMA is still quite a distant prospect. The challenge is not in developing and updating a flight plan to work out how to fly to a new target. That is not a difficult task and can be solved using some quite simple mathematical algorithms that can be readily programmed into the on-board computer systems. Where the real challenges for automation lie is in the area of analyzing and recognizing aberrant patterns of behaviour in images derived from the sensor payload.

Having the ability to automatically decide on a change of target, say if a primary target is covered by clouds or partially obscured, or recognize the kind of activity on the ground that might be indicative of the presence of an enemy, is a major departure from current technological capabilities. In spite of the grave warnings emerging from media sources it is highly unlikely that the man-in-the-loop operation of armed UMA is going to change in the foreseeable future. Even then, one has to question what benefits removing the man from the decision-making process would bring from an operational viewpoint.

Civilian applications for UMA also provide another dimension to their future evolution. Already UMA are used in specialist roles to monitor large outbreaks of wildfires and to patrol sensitive border areas. Disclosures by the Obama administration in 2013 also revealed the extent to which military technologies were now being used in civilian policing operations. This has caused a renewed set of concerns to be expressed about the degree to which homeland defence and upstream military operations are converging.

There is a fine line between conducting operations to monitor the border between Mexico and the United States and then roaming inland to track fugitives as they try to escape from customs and border officials. After all, the techniques developed in Iraq and Afghanistan using UMA to track the activities of specific insurgents could so easily be applied in the case of homeland defence. Taking the worries being expressed in the media to the limit, one does have to ask the question: if you can launch a Hellfire missile onto a moving target in Pakistan or the Yemen, what would prevent someone doing that in America or the United Kingdom?

It is perhaps cruel to suggest it but it would take the considerations of collateral damage to a new level; and yet the idea of killing American citizens using UMA is not new. Anwar al-Awlaki was an American citizen who was targeted in the Yemen because of his known associations with Al Qaeda. If an armed UMA had tracked a known terrorist group taking part in a mountain training exercise in the Rocky Mountains and an opportunity to attack them arose, would someone in the FBI be prepared to sanction it? What about people involved in smuggling drugs? Would they be a legitimate target?

While recent disclosures suggest that the use of UMA by FBI operatives has up until the middle of 2013 been confined to a small number of cases, the concern is that without proper supervision this could rapidly escalate. After all, once the capabilities of armed UMA were fully appreciated the rate at which they were deployed and used grew nothing far short of exponentially.

Other applications will no doubt follow. One of the more exoteric concerns plans in Germany to monitor people who paint graffiti on property owned by the German railway. It costs around $10 million a year to clean, and images derived from UMA are seen as one way to prosecute those involved. In another interesting application UMA have also been deployed in the war on poachers of white rhinos in South Africa. Remaining airborne for up to ninety minutes, the lightweight Falcon UMA can provide important aerial views of suspicious activity.

Proliferation
Another aspect of the future will be the number of actors that are going to be able to fly UMA. The technology of UMA will not just be restricted to nation-states. Already on several occasions Hezbollah has flown UMA over Israel. In early 2013 Iran demonstrated its *Shahed*-129 (Witness-129) UMA. They claim it has a range of over 2,000 kilometres (1,240 miles) and is able to be armed with bombs and missiles. If Iranian propaganda is to be believed – and there are occasions when doubts have been raised over their claims – the UMA is capable of reaching Israel. Its claimed range is twice that previously achieved by Iranian UMA. The development is clearly in response to the increasing surveillance over Iran aimed at monitoring its nuclear programme.

Iran also has a track record of reverse-engineering technologies in order to overcome trade embargoes enforced by the United Nations to try to curtail its nuclear activities. In December 2011 the crash of an RQ-170 Sentinel UMA in eastern Iran led to it being placed on display for the world's media in Tehran. Its stealth-like shape provides an indication of one of the clear directions of travel for UMA technologies.

This was the UMA that had been reported providing imagery over the house in Pakistan where Osama bin Laden was living. Its shape resembled that of a smaller version of the B-2 Spirit bomber deployed by the United States Air Force. Both are based on the idea of a flying wing configuration. Other details on its performance are less clear, although sources in the open media suggest that it has an operating ceiling of 15,240 metres (50,000 feet) and has only been manufactured in similar quantities to the B-2 Spirit bomber. From informal comments sourced in the Pentagon there are currently no plans to arm the RQ-170.

The final frontier?

Space has often been termed the final frontier, picking up the theme from a hugely popular television series that also saw a series of successful films produced in Hollywood. *Star Trek*, however, was based upon the adventures of men as they travelled to the far reaches of the cosmos using warp speed to skip huge distances in a matter of microseconds. Fiction, however, does not always reflect reality. On 11 December 2012 the United States launched the X-37B military space plane on its third highly-classified mission. While it was hardly going where no man had gone before, the purpose of the series of test flights has been shrouded in secrecy.

To an ill-informed outsider the pictures of the X-37B bear a striking resemblance to the now-retired Space Shuttle. The space-plane, as it has been dubbed, is a derivative of a previous development known as the Boeing X-40. It is a re-useable space vehicle that can carry a variety of payloads into low earth orbit. Its first mission was launched on an Atlas rocket from Cape Canaveral in Florida on 22 April 2010. Amateur space observers quickly built a model of the orbit of the spacecraft which they believed was inclined at 39.99 degrees and ensured the vehicle would pass over the same location on the Earth every four days.

Speculation as to the nature of the missions has inevitably been rife in the media. It is an unmanned craft and its payload bay would be ideal to host the kind of radar imaging sensor systems that occasionally flew for a small number of days on the Space Shuttle. It could also be used to house a laser system that could potentially attack military satellites in low earth orbit, although the power required to operate such a system may be beyond the current payload weight and size constraints of the vehicle. The Pentagon has denied claims that this is the real purpose of the X-37B. Another spurious claim in the media suggested that the X-37B was flown to spy on the Chinese Tiangong-1 space module which is to form the heart of their future space-station capability. This was quickly refuted on the basis of the differing orbital pattern.

After 224 days in orbit the first test vehicle de-orbited and conducted America's first autonomous landing onto a runway from space. The first time this had been achieved anywhere had occurred when the Soviets landed their Buran Space Shuttle in 1988. The second mission lasted for 469 days. At the time a Pentagon spokesman noted that the post-mission analysis of the first flight had suggested that it could have remained in orbit for an extended period. One of the aims of the second flight was to push that boundary and see what might be achieved. At the time of writing, the third mission has yet to conclude.

While the mission details of such flights are less than transparent, rumours will always develop to fill the gap. No doubt several more imaginative suggestions will appear in the media at some point. One thing, however, is clear. Having invested so much in the development of UMA to fly in the Earth's atmosphere, to explore their potential role in space makes eminent sense.

CHAPTER 8

Conclusions

Historical perspectives

The early developments of UMA did not arise from a solid military requirement. These were motivated by advances in technology that offered the prospect of a UMA being developed. What anyone was going to do with it did not matter. In modern parlance this was a technology-driven approach. With mankind having only just taken to the air, for many the idea of quickly removing him from the cockpit seemed absurd to those who had only just experienced the thrill of overcoming the effects of gravity using powered flight. The joy was in flying, not sitting on the ground controlling a machine from some distance. Besides, some argued, what could an aerial torpedo achieve militarily that artillery could not already deliver? This was a difficult question to answer.

The problems of flying a UMA were technically demanding. As a subject it was bound to attract the attention of lively minds. Accuracy was a particular challenge but that was not all. If the argument was to be made that an aerial torpedo was to have a meaningful military role, it had to somehow either complement artillery or in some way offer new capabilities. That would mean the UMA either had to offer advantages in range, timeliness of response, variations in warhead or more accurate targeting. By comparison with their contemporaries, these were areas where technological limitations restricted the designers.

The first problem of how to get an aircraft into the air without a pilot at the controls was overcome relatively quickly, although some newsreel clips taken at the time do show that this was far from a trivial exercise. Flight times of more than a few seconds were something of a novelty. On paper the aerial torpedo might have seemed a good idea. In practice it was proving harder to deliver as an operational concept.

Gradually the issues of getting sufficient speed to lift into the air were solved. Once airborne, however, the issue was how to fly straight and level to a notional target to deliver a weapon. In the ground environment the targets on the battlefields of the First World War were relatively static. Nevertheless, weather restrictions could hamper operations. In the air, however, the inherent mobility of the Zeppelins posed a different problem.

Developments in gyroscopic systems provided answers to this problem. As the gyroscope sensed movement caused by wind or other atmospheric disturbances, signals would be generated providing a control input to the elevators and rudder. This would allow the UMA to maintain a heading and a given altitude but over time errors were bound to creep in, reducing the accuracy of the platform. Where a line of sight could be created to the machine, radio signals could be used to compensate for this but even that did not produce a major breakthrough. Gradually ranges over which some degree of control increased as designs and developments of gyroscopes and flight control systems developed. Yet this still did not provide a clear answer to the question as to what military value the UMA actually contributed.

Accuracy therefore remained a major issue. Had the early flight variants ever really been deployed in anger, they would have been little more than a highly-indiscriminate weapon; one of the most important concerns being that the airframe should actually manage to cross the allied lines to attack the enemy.

The Zeppelins

However, as the Zeppelins appeared over London in the First World War, it became clear that something had to be done to try to shoot them down. The sheer size of these behemoths had a psychological impact upon the population. Nighttime operations added another dimension to the fear experienced by the British people. The drone of the engine meant that an attack was imminent: the question was where? In some ways this impact was to be mirrored years later in the Second World War with the onset of the V-1 assault on London.

Apart from flying above the Zeppelins and dropping bombs on them, they proved quite difficult to shoot down. Approaching a heavily-armed Zeppelin in a small fighter only equipped with meagre armaments was also far from risk-free. The resilience of the airships to attack led many to speculate about their design and the kind of gas used in their inflation. Further developments in incendiary armaments were needed before the true vulnerabilities of the airships were exposed.

This was the first time the military requirement for a UMA was apparent. The idea of an unmanned aerial torpedo was an obvious response. However, it had to be effective against a moving target, albeit a slow-moving one. Any unmanned aircraft would have to be steered onto the target. The size of the airships helped but without some form of external control it was inevitable that any aerial torpedo attack would be almost certain to fail. This mandated that any aerial torpedo had to be controlled either from the ground or by a pursuit aircraft flying nearby but out of range of any defences on the Zeppelin.

In the wilderness
The nascent stage of development of key technologies in navigation, flight control systems and radio communications did provide significant barriers to their use in the First World War. Of the three, developments in radio and flight control systems were to quickly advance to the extent that at least some degree of remote control could be established over a UMA. In the austere times that existed after the war, further developments in UMA-enabling technologies, such as in the field of navigation, did not provide a huge magnet for limited research and development funds. Many programmes became dormant through lack of investment.

The core enabling technologies, however, did manage to attract research funds and significant developments began to occur. Once they were developed in sizes that made them readily applicable to being installed on aircraft, the next round of UMA developments was almost bound to follow. Many of these technical advances were spurred on by rapid developments in civil aviation in the late 1920s and early 1930s. Imperial Airways were in the forefront of opening up new routes across the world, flying from the United Kingdom to the far-flung corners of the Empire.

As civilian aircraft spread their wings across the world, new military aircraft also started to appear. Some of these were specifically designed to dive-bomb their targets. Up until then this tactic had been seen to be very risky but with the introduction of aircraft carriers the need to be able to increase the accuracy of delivery of a weapon against a warship became increasingly clear. For navies, defence against such a threat required new forms of naval gunnery.

The first serious military application of UMA was therefore as target drones for naval gunnery training. Previous targets had been rather predictable in their motion. Once control could be established over the UMA to the extent that it could be made to dive at shallow angles over a target, their use increased. In many ways the early versions of what became known as target drones were ahead of the kind of tactics that would become commonplace in the Second World War when it came to dive-bombing enemy warships.

The early pioneers of UMA had focused their thinking on using such a device to deliver explosives against enemy targets. This followed what might be viewed as a traditional approach to the ways in which kinetic air power was applied. It was all about delivering bombs against targets. The Austrians had started this line of thinking in the Siege of Venice of 1849 but when balloons were flown over the city they were only challenged by relatively inaccurate ground fire.

As air-to-air combat quickly developed over the stalemate of the military position on the ground in the First World War, the survival time of bomber crews reduced. Having a remotely-flown aircraft at least did not expose air-crews to harm. Accuracy of delivery was, however, a problem as the weapon could quite literally land anywhere. For some military people this unreliability was an issue but for others it created an opportunity.

The V-1 campaign

The V-1 campaign against London in the summer of 1944 was the first time cruise missiles were employed in a significant role in a military conflict. The air component of the attacks launched from Heinkel-111 H-22 bombers was totally ineffectual. The losses of air-crews hardly justified the impact the air-launched weapons had on the population of the United Kingdom. The ground-based missiles proved more effective as a terror weapon. For the population of London who had endured the Blitz, the onset of the V-1 campaign provided a clear reminder that the war was far from won. However, their reaction to the attacks was inevitable: they soldiered on.

Since the inception of air power a population has only once been cowed by air power and that involved the use of nuclear weapons. By contrast, Adolf Hitler's view that his so-called vengeance weapons would force the Allies to negotiate proved incorrect. Even had the launch rate approached the levels that Hitler foolishly believed possible, it is doubtful that the V-1 campaign would have created a tipping-point in the war.

History would no doubt have been completely re-written had the Nazis managed to develop a nuclear weapon and install that on the V-2. As the war drew to a close there is little doubt the Nazis would have used such a device on London. If anyone would like to doubt that assertion, just consider the onslaught that was unleashed towards the end of the war on Antwerp. The death toll of Allied servicemen and civilians in the port and its surrounds is often forgotten in discussions on the impact of the V-1 weapon.

CONCLUSIONS

Throughout the V-1 attacks the Nazis lived in the same self-delusory world that many of them had occupied during the Battle of Britain. Many were all too ready to believe the intelligence assessments they were receiving. Their own propaganda reinforced the prevailing *zeitgeist* that they were winning. The Nazis simply had little idea of the increasing effectiveness of the defences over England. In part this was down to some effective cooperation with the media in the United Kingdom. When the very existence of a country is at risk it seems even the media can temper their desire to publish a good story.

As during the Battle of Britain, the measures taken to defend the country against the V-1 attacks were adapted to cater for the emergence of a new threat. Fighter-pilot tactics evolved, as did the ways in which the multiple layers of defence provided by the anti-aircraft guns, the balloons and fighters interacted. Each was able to understand, as dictated by the prevailing weather conditions, the best way to run the battle on a day-to-day basis.

Unfortunately this level of organization was not something that could be moved to protect Antwerp. Unlike modern defences against missiles, such as the Israeli Iron Dome or the American Patriot systems, the defence that was created against the V-1 for the United Kingdom was not easily relocated over Antwerp. The suffering of that city provides a vivid demonstration of the scale of what London might have faced had the defences not proved so effective.

Had the V-1 campaign against the United Kingdom been sustained, it is likely that it would have reached a kind of stalemate with a small percentage of missiles still reaching London and the south-east from the fixed launch sites in France. Given the Allied bombing raids in Germany and against the launch sites, it is difficult to see how the Nazis could have surged the level of attacks to a point where the air defence systems would have been overwhelmed.

Equally, not all the V-1s launched could have been intercepted, so over a period of time the rate of successful attacks would have declined irrespective of the impact of Allied forces sweeping across Europe. The airborne component was never likely to grow significantly and the Mosquitos were taking a heavy toll on the He-111 H-2s from which they were launched. Over time, that component would have diminished still further.

With the end of the Second World War, applications of UMA returned to their traditional base of being targets for gunnery practice. Even today that continues as they are used to test the effectiveness of air-to-air and surface-to-air missiles. It will remain an enduring application of UMA, with their other long-standing role being that of reconnaissance.

As the V-1s started to rain down on London in 1944, in the United States pioneering work was just starting on the development of a pilotless aircraft designed to attack targets deep inside Germany. One motivation for this research would undoubtedly have been the sheer scale of losses of daytime air-crews suffered by the Americans. Again those tests were conducted at a point where the technology had not advanced sufficiently for the concept to ever really get to the point where it could be operationally deployed. Only thirteen unsuccessful test flights occurred before the programme was abandoned but it had given another stimulus to the progressive development of UMA technologies. Slowly, step by step, investments were being made to allow the challenges of unmanned flight to be overcome.

The SEAD mission
If there was to be a catalytic point at which UMA would enter the mainstream of military applications it was always likely to occur during the Vietnam War. As flying over Vietnam became increasingly hazardous, so the idea of using UMA to suppress the effectiveness of Viet Cong defences developed. UMA would be flown in a stand-off mode to broadcast jamming signals to disrupt the performance of its increasingly capable air defence systems. The SAM systems employed by the Viet Cong depended upon radar systems for their initial target acquisition. What evolved over Vietnam was the second generation of electronic warfare. The first had occurred during the Second World War as the Nazis and the Allies battled it out for supremacy of the radio spectrum. The whole notion of measure versus countermeasure started the first of numerous spirals in the Second World War.

This use of UMA to analyze the electronic ORBAT of an adversary's radar systems should have provided an enduring legacy from the Vietnam War. However, as radar technology improved so the collection and analysis of SIGINT data became more complicated. The sheer variety of radar waveforms and modes that started to appear made analysis of radar systems a task that was not easily automated.

For those involved in the development of RWR handling, the uncertainty of the electronic warfare environment was a challenge until the development of advanced digital signal processing. Analyzing SIGINT data required specialist knowledge and this was not easily transferred into a piece of software that could make automatic decisions on how best to jam enemy radar systems. Manned systems were the obvious way forward for SEAD missions. The manned SEAD mission provided the main electronic warfare component over Iraq in 1991 and 2003.

176

CONCLUSIONS

The Israelis took a different view. The lessons from the Yom Kippur War had been harsh. The initial rate of aircraft losses over the Sinai Desert and the Golan Heights had almost taken Israel to the edge of the abyss. In the crucible of war one event occurred that was to have a major impact on the ways that UMA would be viewed in the future. The arrival of thirty-three Lightning Bugs from the United States during Operation NICKEL GRASS created a dynamic that was to fundamentally change perceptions over the applications for UMA. Once the war was over, the Israelis were to pioneer several new developments of UMA technology that would lead to the development of contemporary systems.

The first indication that this strategy was working came in the Bekkar Valley in 1982. Confronted by a highly-effective Syrian Air Force equipped with contemporary Soviet warplanes supported by an air defence system, the Israelis showed how they had mastered new roles for UMA in jamming communications and radar systems. Despite the obvious success of the SEAD mission over the Bekkar Valley, it had little impact on developments in the west. If anything, the requirement for the SEAD mission for UMA began to fall away.

What had proved so successful tactically from a manned electronic warfare perspective over Iraq in 1991 was to have limited effect in Kosovo less than a decade later. Since then there has hardly been a massive call for the SEAD mission in places such as Afghanistan and Mali. Where air forces are essentially fighting an asymmetric conflict there is little need for SEAD and in the Libyan campaign in 2011 the age of the air defence system operated by Colonel Gaddafi's forces did not require a full-blown SEAD response. SIGINT platforms like the Nimrod R1 were used to collect intelligence on the Libyan air defence system but much of its operational characteristics were already known. In the first four days of the campaign the NATO-led coalition degraded the operational capabilities of the Libyan air defence system to a point where it was considered almost ineffective.

These recent experiences, however, do not point to a world where the SEAD mission will no longer form part of the repertoire of air force operations. Conflict over key parts of the South China Sea could draw America and China into war. Trying to attack key mobile assets operated by the Chinese that would be high on the targeting list of any American carrier-based operations in the region would require a SEAD response. This echoes what happened in Vietnam when UMA were equipped at relatively short notice to conduct SIGINT missions that were then able to help devise suitable responses, such as barrage jamming.

While electronic warfare can be implemented using brute force, there are other times when a more subtle approach is required. Where brute force is needed the power requirements are often met by manned aviation escorting packages of aircraft to a target. Where a more nuanced approach is required UMA can be deployed that exhibit similar behavioural characteristics to manned aircraft with a view to seducing the enemy into revealing their wartime radar operating modes.

On a number of other occasions they have also been used to decoy surface-to-air missile systems away from manned targets. These implementations of UMA were little more than an advanced form of target drones that originally started to be built in the late 1920s. They were essentially expendable. Their missions were also often of a relatively short duration. Reconnaissance missions, however, were different. It was in that area of UMA development where mission duration and the issue of persistence were to be major drivers for new advances in UMA technologies.

Yom Kippur

The Malthus doctrine was 'Necessity is the mother of invention.' In Israel the difficulties of fighting wars in both rural (Lebanon) and dense urban (Gaza Strip) environments led to some important advances in UMA technologies. When it came to conducting military operations the Israelis were not immune from the voice of international opinion.

The need for delivering more precise effects became a necessity. Pictures of dead and injured Palestinian children did not play well in the international media but Israel could hardly sit back and allow HAMAS and Hezbollah to launch rockets against their citizens without some form of reprisal. Therefore a twin-track approach was adopted. Senior players in the organizations behind the attacks would be targeted, as well as those launching the rockets.

To achieve this Israel had to pioneer new developments in persistent UMA. This required advances in propulsion systems and miniaturization of the basic components required to fly and operate a UMA. Fighter jets were simply too expensive a solution. What was needed was a smaller, longer duration platform that could be equipped with advanced sensor systems to identify and track potential targets before they were handed off to strike aircraft for interdiction.

Individuals had to be tracked over a period of time before the right location was identified to conduct a strike operation. Images of burning cars in the streets of Gaza provide a demonstration of the increasing accuracy with which such missions are carried out. However, there are times when this goes badly wrong. Those quickly erecting a rocket-launcher to fire an unguided missile into Israel are also a fleeting target that could only be addressed by having persistent coverage of likely launch sites. These urgent military requirements, coupled with a political necessity to be shown to apply force in moderation, led to the development of a new generation of UMA.

UMA also have advantages over satellites. Current technological limitations on sensor systems restrict access to high-resolution imagery from geo-stationary orbit. Low earth-orbiting satellites follow predictive paths that nation-states can predict, allowing countermeasures to be adopted that reduce the value of any intelligence derived from space. UMA, however, can persist over an area for a long period of time, helping to develop the strategic picture of what is happening on the ground.

America's use of UMA against the Iranian nuclear programme has been well covered in the media. This is the modern-day equivalent of flying UMA over China and areas of the Soviet Union in the Cold War. The use of UMA helps avoid the kind of internationally embarrassing incident that occurs when the airmen flying them are captured and paraded on television.

The IED threat

Malthusian thinking also permeated to coalition forces operating in Iraq and Afghanistan. When faced by an increasing IED threat the necessity was to find ways of reducing the increasing death toll among the military forces based in those two troubled countries. The main problem was one of persistence. To spot someone planting an IED on a known route used by coalition forces, a sensor system had to be present. That could either be on the ground or in the air. The sheer magnitude of the task of deploying and maintaining a sophisticated network of ground-based sensor systems or the risks associated with basing servicemen on the ground to watch major re-supply routes required an airborne solution.

Fortunately developments in electronics, airframe technologies (lightweight composites) and power plants provided the catalyst for a new genre of UMA to appear. This iteration was able to make some limited decisions of its own if it lost contact with its home base. This was the point at which the uses of UMA multiplied dramatically. Other forms of smaller, lightweight hand-launched UMA started to appear. These were driven by real operational needs in theatre.

One of the unsaid reasons for their introduction was the aversion to military casualties that now exists in western society. Media coverage of repatriation ceremonies creates uncertainty in the minds of the public, who often are unsure why servicemen are being sent to far-off places to fight wars. Political leaders in the west have failed on many occasions to provide a telling case justifying upstream military interventions. This public concern translates into a desire to see the numbers of troops killed and injured being kept to a minimum. Any piece of military equipment that could be seen to help reduce the casualty count simply had to be developed and deployed if it were technically feasible.

This creates problems for governments over defence equipment procurement. The UOR became the way in which all sorts of new developments were rushed into Iraq and Afghanistan. However, a legal judgement in June 2013 after servicemen had been killed in what were regarded as vehicles with limited armour protection has the potential to dramatically change the ways in which equipment is purchased for future wars.

Threat assessments will have to operate outside the kind of tight box in which they have traditionally been contained. While empirical data is yet to be available, there are clear patterns emerging that suggest the proliferation of new weapon technologies is occurring at a faster rate among transnational criminals and terrorist organizations.

What this means is that any military venture, no matter where in the world the forces are deployed, will face an increasingly sophisticated threat from the enemy. The tactics of asymmetric warfare used by the insurgents in Iraq and Afghanistan have escaped like a genie from the lamp but on this occasion it appears unlikely that it will ever be returned. In such environments UMA are a vital element in helping the overall situational awareness of servicemen on the ground. As a result of the legal judgement in the United Kingdom, having such equipment is now their human right.

In Iraq and Afghanistan UMA appeared in a number of new designs but the essential building-blocks remained the same. Flight control systems, communications equipment and a payload of sensors were the very minimum elements of a UMA. For those in service at an operational level of command it was also sensible to arm them in order to reduce the sensor-to-shooter time to a minimum.

Small tactical UMA appeared that could 'look around the corner' in a dense and complex urban environment. They could also be flown ahead of a convoy to look for telltale signs on the ground of insurgent activity. At the operational level UMA were also used to provide intelligence on the activities of specific individuals or groups. The UMA proved adept at tracking their activities and providing indications of aberrant behaviour.

CONCLUSIONS

The human factors

Among all the negative press about UMA, one factor is often overlooked. What toll do repeated operations take on the pilots and weapon systems operators who fly the missions and conduct attacks? One of the classic arguments the media like to portray is that UMA warfare can be likened to computer games. The problem with this analogy is that computer games rarely have the endless hours of boredom that are experienced by UMA operators. Computer games tend to be all action. In the real world of UMA operators there are hours upon hours of simply watching potential targets up to the point where a strike may be carried out while minimizing the risks of collateral damage.

This is a perception fuelled by the notion of UMA being flown by people who are good at this kind of virtual warfare. The suggestion is that these individuals lack the kind of psychological connection with warfare that can only arise from being in a theatre of war. The argument goes that unless as an individual you are up close and personal with the enemy, you cannot believe you are at war.

This, of course, is a very biased viewpoint. Few professional military people would accept that flying UMA from an air-conditioned office in the Nevada desert somehow makes you immune from the realities of war. Indeed, such is the scale of integration that the UMA operators are in contact with the combat units they are supporting on the ground. Listening to the radios often provides a very clear indication of how pressured the ground troops feel in specific situations. The UMA operators cannot help but be drawn into the situation when they hear a fire-fight in progress.

Despite the media suggesting otherwise, the UMA operators do not see their missions through the lens of a video game. Recent studies show that the reality of the situation is far more complex. What in military jargon is known as decompression, the point at which a serving person leaves an operational theatre to resume their home life, is very hard to do on a day-to-day basis.

Those who fly UMA from a safe and secure sanctuary in the United States or the United Kingdom and then drive home to put the children to bed suffer different forms of psychological pressures. One minute they are listening to calls for help from soldiers on the ground who have been ambushed and the next they are reading a bedtime story to their children. It is nonsense to suggest that such episodes of diurnal decompression do not have a lasting impact on the operators. Developing coping strategies in such situations can often require medical help.

The operators also suffer from another level of exposure to what is happening on the ground. Today's sensor systems leave little doubt as to what has happened in the aftermath of an attack. Fast jet pilots will fly over a target, drop a weapon and then retire from the scene, leaving others to do the bomb-damage assessment. For the UMA operator the implications of their attacks are all too clear to them. When mistakes do occur, the results are all too graphic. This piles yet more pressure on the human operators.

Media portrayals of the use of UMA being little short of a video game are certainly far from reality. They are also insulting to the professionals who operate the systems. Reports of increasing levels of Post-Traumatic Stress Disorder among UMA operators back up that viewpoint. It would appear that far from living their lives in some virtual world, the UMA operators really do experience high levels of stress. These effects are likely to be amplified by the operational tempo that the current generation of UMA operators are sustaining. Due to a slow build-up in the total numbers of air-crew, those who are qualified are carrying a heavy load. That is bound to reveal itself psychologically at some point.

Contemporary operational issues

Today UMA get a bad press. They are either automatic killers raining death and destruction from on high against war-weary insurgents in Pakistan or they are intruding upon people's personal lives at home. The times when they are used to help human beings are often barely noted by the media. Even their supposed cheapness is questioned. For many, some of whom have clear agendas, UMA are not some kind of panacea.

The loan by Israel of two UMA to the Chilean government in the aftermath of an earthquake that had widespread impact upon the population was hardly reported. For the Chilean government the ability to assess those areas that had been most affected and to help provide security at nighttime when the risk of looting was high was a humanitarian gesture. This was one occasion, and there have been many others, when UMA could have been noted for their positive contribution to helping human beings.

Another example is the way in which UMA are being used to chart ancient Inca settlements in Peru. Their use speeds up the rate at which initial surveys of sites can be completed. For the Peruvian economy this is important as government statistics show that developmental pressures have surpassed looting as the major threat to the country's national heritage. At a site near Lima archaeologists are trying to piece together the remains of a fire-revering coastal society that was nearly destroyed by construction firms in July 2013. Therefore UMA clearly also have a role in conservation.

CONCLUSIONS

In addition, UMA are increasingly being sent into areas that are very dangerous places for human operations. For example, UMA are being flown into hurricanes to measure and report atmospheric parameters. They have also been used to chart the plume from an erupting volcano.

Several universities are engaged in using UMA to monitor levels of sea ice that might provide indications of global warming. Scientists at the University of Alaska at Fairbanks are also using a small battery-operated UMA to map the summer breeding grounds of sea lions. Forest and grassland fires are also being monitored using UMA. These are all positive aspects of their use. Not all UMA are armed and dangerous. Some make a very positive contribution to society.

Negative press
As ever the media fails to find any sort of balance in the treatment of the subject. What started out in 2006 as claims of large numbers of civilians getting killed by armed UMA strikes could never quite be reversed, despite the existence of evidence that suggested the civilian death toll was lower than the media hubris had suggested.

Publish by exception is the rule. Only mention the topic when many people have died in an attack, irrespective of their allegiance and possible involvement in international terrorism. Stories of the day-to-day roles played by UMA in securing countries' borders, tracking drug-smugglers and preventing IED attacks upon military personnel based in a theatre of war are consigned to the editorial waste bin. They simply do not suit the agenda.

It seems that the public are easily swayed by such coverage. Reticence over the operations of UMA often appears in opinion polling across the world. Women are found to have an especially negative reaction to the armed UMA strikes. Doubts expressed in some sections of the media over the legitimacy of the operations from a legal standpoint add to that sense of disquiet. Protests outside Royal Air Force Waddington over the operations of UK UMA over Afghanistan provide an indication of the kind of concerns that exist. The images of charred bodies after a Hellfire missile attack easily stir public unease.

Reassuring noises emerging from Washington over the increased level of controls involved in targeting decisions do little to assuage the *zeitgeist*. Public opinion, it would seem, is far from convinced about the idea of a pilot and weapon systems operator being detached from the point of weapon delivery. They want the pilot to be up close and personal with the war, failing to appreciate that thanks to technology, that is exactly where they are when they are flying overwatch missions. The real failure here is on the part of those who operate the armed UMA. They have allowed the media image to become tarnished by dubious reporting.

The problem is that the media allows what happens in strikes aimed at the Taliban and surviving leadership of Al Qaeda to be conflated with the need for soldiers on the ground to be protected. A nuanced coverage is absent from a large section of the media that is hostile to the very idea that UMA exist. It is difficult to completely understand the rationale or motivations behind this attitude. Soon this may be compounded when the numbers of people killed by armed UMA strikes exceeds the total number of people who died on 9/11. When this milestone is passed, will any public support for further armed UMA strikes ebb away?

While UMA suffer from adverse media attention, from a military viewpoint they make increasing sense on the complex battlefields of Iraq, Afghanistan and in tackling transnational international terrorist gangs. The French operations in Mali in 2013 provide an ideal example of the problems military commanders now face in trying to seek and destroy what is often a fleeting enemy. Until the French intervention in January with the help of the Royal Air Force and the Americans, Al Qaeda-affiliated groups in Mali had been able to operate without fear of sanction for their activities. Historical religious sites were desecrated in Timbuktu, meaning that some material is now lost from the annals of history. The world appeared impotent as these groups mounted an onslaught against those who practice rival interpretations of their religion.

When it appeared the wider international community was faltering in its desire to finally try to stabilize the northern areas of Mali, the terrorists moved south in an audacious move to take the capital Bamako. French fighter jets supported by UMA and ground forces helped to quickly remove that threat. However, as is inevitable in the modern era of warfare, the insurgents did not stand and fight. They retreated into the Adrar des Ifoghas mountain range which borders on Niger, Libya and Algeria.

To many this appeared like a re-run of the sanctuaries created by Al Qaeda and the Taliban in the unregulated areas that exist along the border between Pakistan and Afghanistan. In Somalia Al Qaeda affiliates operate over large swathes of the countryside and a similar situation exists in the Yemen. These ungoverned spaces, as they are known, provide the perfect sanctuary for transnational criminal and terrorist groups.

As the French military operations in Mali started to scale back it seemed that rather than achieve a decisive military victory against the insurgents, all they had actually managed to do was disperse them over a wider area of the Sahel. Reports of Al Qaeda-related activity now routinely emerge from Mauritania, Western Sahara, Morocco, Algeria, Tunisia, Libya, Niger, Chad and Mali. For any land-based military forces, especially those indigenous forces that are so poorly equipped, trying to create any form of governance over such a huge and often inhospitable area is a massive task. Armed and unarmed UMA flying in these areas represent an obvious way of trying to create a semblance of governance and presence. Just the sound of the engines of the armed UMA still creates its own impact on those on the ground.

Far from defeating the insurgents, the French military operation has ensured that yet more sanctuaries will now emerge which the groups involved will use as yet another base from which to start future activities. How to maintain pressure on those locations while achieving the military draw-down to which the government has committed poses a problem for the French. One indication of the long-term prognosis in Paris is the decision to buy at least five MQ-8 Reaper aircraft for the French Air Force.

This is a significant development and reveals how the Élysée Palace is covering what may happen next in Mali. The Russian decision to buy Israeli armed UMA mirrors the reasoning in France. While both governments would prefer to await their own indigenously-developed armed UMA, they both have pressing needs on the ground to address insurgencies that threaten their national security. Terrorist training camps in the Caucasus, across the Sahel, Maghreb and Middle East are, like those in Somalia, Nigeria, Pakistan, Syria and Iraq, an enduring concern. The inherent manoeuvrability that Al Qaeda franchises now enjoy has led to the approach of using armed UMA strikes to try to disrupt their activities being likened to 'whacking molehills'. It is a very good analogy.

Recruits into those training camps will also flow from the spate of prison escapes that were engineered towards the end of July 2013. From supposed secure sanctuaries in Iraq, Pakistan and Libya hundreds of former combatants have escaped. For President Obama the idea of repatriating the remaining people held at Guantanamo Bay has become even more difficult. In the Yemen a mass prison escape saw hundreds of Al Qaeda supporters released to join the ranks of AQAP.

With more of these people now travelling to far-flung parts of the Middle East and North Africa where they can resume their jihad, the international security situation has taken a major step backwards; and this is without any additional potential fall-out from the ousting of President Morsi in Egypt. With Al Qaeda-linked groups already operating in the Sinai Desert, how long will it be before Israeli armed UMA are involved in strikes against training camps in that area?

In this uncertain environment UMA provide almost the ideal platform from which to monitor and where necessary disrupt the activities of the remnants of Al Qaeda's franchises and affiliated groups that have established bases in ungoverned spaces. Some commentators speak of Al Qaeda metastasizing, using a metaphor drawn from cancer research to convey the idea that new cancers develop away from the primary source. Where these new cancers metastasize to new geographic areas, some form of upstream action has to follow.

UMA provide the unique quality of persistence over the potential target areas that are so valuable. They provide a cost-effective alternative to what would be resource-intensive and expensive-to-maintain manned missions. It is unlikely to be long before reports emerge of UMA strikes against insurgent leaders who have based themselves in these countries. The decision by the United States to establish a forward operating base for UMA in Niger in the early part of 2013 shows how the footprint of UMA is gradually extending across Africa. Where Al Qaeda and its affiliates go, UMA such as the Reapers will inevitably follow. That is not the only development that is occurring with regard to UMA.

Understanding the public reaction to those occasions when civilian casualties die during armed UMA strikes, President Obama has sought to codify the rules of engagement into a clearer set of processes for analysis and decision-making. Suggestions emerging from the White House note that this has been important as armed UMA strikes are likely to remain on the agenda for some time. Some commentators close to the Obama administration have privately noted that the so-called 'global war on terror' is only at its halfway point in 2013. Clearly there are few in the administration who believe the utility of armed UMA strikes is suddenly going to fall away.

In this situation the armed UMA strikes are now supported by the development of what has been called the Disposition Matrix. For each person on the target list it contains the list of measures that are being taken against the individual involved. Some may be subjected to arrest warrants and other legal actions; others to actions by Special Forces, especially if they are out of reach of the armed UMA capability. The approach based upon this new matrix provides greater clarity for those involved in the decision-making processes.

The decision to send in two Special Forces teams to arrest key people linked to Al Qaeda in Libya and Somalia in the wake of the assault on the

CONCLUSIONS

Westgate facility showed the full potential of the Disposition Matrix. One of the raids was successful; the other failed to achieve its objective. Simply killing key leaders as soon as they come into the cross wires of an armed Reaper does have its consequences. Clearly the White House, in its attempt to rebut some of the criticisms that have been laid at its door over its apparent widespread use of UMA, has decided to look more carefully at other options offered by the Disposition Matrix analysis.

This gradual shift that can be detected in 2013 comes at a critical time as Al Qaeda spreads its global footprint. Its increasing geographic coverage was a point noted by its leader Dr Ayman al-Zawahiri in a broadcast in November 2013 where he trumpeted the increased global footprint as an example of the enduring presence of Al Qaeda on the international security landscape. With Al Qaeda far from defeated and many new potential sanctuaries opening up across North Africa, the pressure to conduct more UMA strikes will increase.

This future growth of UMA strikes has already been forecast in some sections of the media. One leader in a major national American newspaper even went so far as to suggest that President Obama has plans to extend the deployment footprint of armed UMA all over the world. No doubt if asked, the president might respond: 'If that's what it takes to keep Americans safe, so be it.' Despite his obvious personal values and ideals, the president knows that holding the reins of power sometimes requires a degree of pragmatism when it comes to decision-making.

Because the targets are often fleeting in nature, the time taken from detecting the threat to any engagement has to be reduced to an absolute minimum. The so-called sensor-to-shooter cycle has to be reduced to a minimum while still allowing the right levels of legal and political overview. Herein lies a continuing problem. To some degree the president can pre-authorize the engagement of specific targets on the list of those thought to be involved in international terrorism. However, each situation has its subtleties. As the list of surviving key leaders is reduced, so the considerations involved in targeting them have to develop.

Cost

Critics of UMA often cite the dubious mathematics that are applied to the ways in which their operating costs are presented. As far as the general public is concerned, it is easy for manufacturers to make the case that UMA cost less to build but the arguments depend upon what type of platform is being analyzed. Clearly those hand-held UMA that have relatively short-duration flying time and are used in a tactical context are low-cost. They are not designed to replace other capabilities. They were introduced in Iraq and Afghanistan as a matter of necessity. Malthusian logic was again at work.

The issues on cost really focus on the capabilities offered by a Reaper UMA versus that of an F-16 fighter jet. On the surface it may look like a really easy case to make, with an F-16 costing anywhere between £5M and £35M depending upon its availability in the second-hand market and the equipment it carries. The unarmed Israeli Heron UMA could be purchased for around £12M in 2004. The armed Reaper price is around £20M. The Global Hawk platform cost NATO around £200M each in the contract signed in 2012. However, while the price of buying such platforms provides a far from clear view of their merits over manned aircraft, a detailed examination of the long-term operational costs and combat loss rates provides a very different perspective. To support Reaper operations a reported 180 people are needed on the ground. By contrast, an F-16 requires 100. Clearly, from an operational viewpoint, if the F-16 were to achieve the levels of persistence of the UMA over the battlefield it would cost a lot more to keep it airborne and being refuelled.

This serves to illustrate the complexity of the actual calculations that need to be made to compare the cost of operating UMA and their equivalent manned fighters. One complicating factor is the relatively high loss rate that is occurring with UMA platforms. This is probably a passing phase. All new technological developments go through the 'bath tub' curve of unreliability in their initial operational deployments but eventually settle down into a low loss rate. Even manned fighter jets went through that stage of their development. In 1954 United States naval aircraft suffered loss rates of close to fifty for every 100,000 flight hours. Today that figure is less than one. As an indication that the reliability of UMA is on a similar pathway, in 2011 the loss rate of Predators fell below that of the F-16.

Ultimately an important factor will be the degree to which sensor systems for both manned and unmanned aircraft develop. If both can achieve higher degrees of situational awareness on the ground to improve precision to the point where the risk of collateral damage is reduced to a very low probability, then the arguments shift onto which platform is best to execute the mission. In uncontested airspace those arguments are likely to favour the unmanned aircraft, even though the support costs on the ground are higher. For manned aircraft to try to compete with UMA in terms of persistence over the target is difficult. In contested airspace, however, for the immediate future the arguments will favour the manned platform.

CONCLUSIONS

Future military applications

UMA will also make an important military contribution in future theatres of conflict. French military operations in Mali in 2013 provide an insight into how the future might evolve. For states that are increasingly thought to be at risk of being unable to govern their remotest areas UMA provide an ideal platform from which a long-term intelligence picture can be assembled. If and when some form of military intervention is required, intelligence collected from surveillance operations will help shape the planning process. With Al Qaeda supporters now spread all over the Sahel and Maghreb, the need to find their new sanctuaries in order to continue to disrupt their activities is increasingly clear.

What emerges from Mali is a new form of military intervention that is an adaptation of the model applied in Libya and Sierra Leone. This is one that sees a short phase where soldiers are deployed on the battlefield alongside forces from the host nation to solve an immediate political crisis. The British intervention in Sierra Leone had similar easy-to-define political objectives.

The Libyan experience was, of course, slightly different in that the footprint of soldiers on the ground from the coalition implementing the United Nations Resolution was kept to a minimum. In the case of Libya, support to the insurgents came principally from the air from overseas bases and from the decks of warships operating close to the Libyan coastline. These are, however, variations on a broader model of military interventions that seeks to reduce the time spent on combat operations to a minimum.

Efforts to build the capabilities of the national army then follow as the intervention force is gradually reduced and returns home. In the gap while the national army is being developed, Special Forces from countries supporting the stabilization efforts move in to ensure that any sanctuaries to which insurgents or guerrillas have temporary retired are kept under pressure. All the while UMA are used to locate and engage the enemy as they try to establish remote sanctuaries in ungoverned areas.

This is the kind of model that is likely to emerge in Mali. However, as events in Syria are showing all too graphically, even this relatively new model for military interventions has its serious limitations. Yet even against that difficult and highly complex backdrop, UMA could still be deployed to feed live imagery to those on the ground that are receiving support. Achieving that is not difficult once operators of the small hand-held terminals that receive data from the UMA have been trained. In the first instance support from the UMA would be through the supply of imagery to operators on the ground.

Operating from Cyprus, UMA could easily spend long periods of time over Syria. If western governments were finally to decide that a red line had been crossed on chemical weapons then arming the UMA could be a logical next step that would avoid any military footprint on the ground. This would be an unmanned variation of the operations over Libya in 2011. The only issue for the governments involved would then be the problem of what to do if one of the UMA was shot down by the Syrian air defence system. Such operations may also increase the risk of terrorist attacks at home. In France the possibility of a reaction to their intervention in Mali will remain a very real threat.

Many leading military thinkers have started to express views on the nature of future wars. Climate change is one driver that many believe will inevitably lead to more conflicts. Where states argue over finite resources, such as water, there is always the potential for tensions to spill over into hostilities. In such situations UMA can even play the role of peace-keeper. The decision by the United Nations on 1 August 2013 to use an unarmed Italian surveillance UMA in the Democratic Republic of Congo is unlikely to be the last time UMA are deployed in support of peace-keeping operations.

To date the United Nations has relied upon individual states contributing to peace-keeping forces bringing their own equipment with them. For operations in Africa, such as those being undertaken in Mali, indigenous forces in Africa do not have access to such advanced technologies. Countries such as France, the United Kingdom and America need to provide UMA as a coalition asset into such operations.

As tensions threaten to flare up into all-out conflict, UMA operating from nearby bases can provide the kind of insights that may help countries on a pathway to war step-back. War often arises from uncertainty, where potential adversaries' understanding of the intelligence picture is often distorted by deception and intrigue. UMA operated by the United Nations could give it specific access to intelligence information that may help create the conditions for negotiations. They can also help verify the implementation of agreements, such as the withdrawal of forces from a potential flash-point.

UMA can also play an important role in helping save lives. Where natural disasters strike, prompt deployment of ISTAR assets can be crucial in directing aid relief efforts. Nearly every week in the press new applications are also emerging all over the world. Monitoring the aftermath of man-made disasters is another area. The nuclear disaster in Japan saw UMA deployed to help build situational awareness. UMA equipped with radiation sensors can go into the kind of environments that are simply hostile to human beings.

CONCLUSIONS

Where force does need to be applied it will be important for the arms industry to develop new programmable warheads. The Israelis showed the benefits of using unarmed warheads to signal the imminence of an attack during Operation CAST LEAD. In the kind of mixed urban and rural environment that will continue to provide the backdrop to armed UMA operations in the immediate future, the ability to scale the size of the explosion is a step worthy of future investment.

In March 2009 the MQ-9 Reaper was first equipped with the 500lb variant of the Joint Direct Attack Munition (JDAM). The platform can also carry the 500lb GBU-12 Paveway II laser-guided bomb to attack specific targets. The GBU-39B small-diameter bomb was also added to the list of weapons that could be carried on the MQ-9 Reaper to make a start in addressing the issues of lowering the risks of collateral damage. In 2010 the MQ-9 Reapers were updated to carry the Hellfire AGM-114 P+ missile. This design allowed the missile to be fired off-axis, overcoming a previous operational restriction. The development of the new missile increased the engagement envelope.

Already the United States navy has made it clear this is an avenue it is actively exploring through the development of the Selectable Output Weapon (SOW). Once the target is known, the pilot selects the explosive power that the device will yield on detonation. Initial trials have already proven the idea. In one experiment a 27-kilo bomb was operated in three different modes. In the deflagration-only mode a mirror located less than 2 metres from the target remained unscathed after an attack.

Where wars do break out UMA may well have to operate in contested airspace. Over Iraq and in Afghanistan the UMA did have to survive ground fire when operating in mountainous terrain but they did not have to deal with the issue of counter-air. Over Kosovo in the late 1990s where UMA first started their current journey fifteen were lost to SAM engagements and door gunners operating from helicopters. Their slow speeds and lack of any form of defence make them unsuitable to operate in areas where high-intensity conflicts are likely to occur.

As America shifts its focus to the Pacific Rim away from its lengthy military commitments to Western Europe and the Middle East, it will have to think hard about how any form of UMA will operate in the kind of non-permissive environments that may arise in that region.

A high-intensity conflict in South-East Asia over oil and gas deposits in the South China Sea is not going to be a place where armed UMA are likely to have a lengthy life expectancy. The emphasis upon stealth in designs such as the Phantom Ray UCAV (X-45) and the X-47B show the likely development trajectory. Other important aspects of survivability will include greater speed and the need to carry a defensive aids suite and perhaps even some form of air-to-air missile capability.

In high-intensity conflicts stealth characteristics are still going to be important. In some quarters in the United States there are already suggestions that armed UMA have had their day. Given the need for these developments some commentators express the view that while they were right for the COIN environment, they simply do not have the wherewithal to operate successfully in a contested environment. The suggestion is that from here on the pressure to further develop armed UMA will recede. In China and Russia indigenous UMA development programmes suggest the thinking in Washington is not shared by Beijing and Moscow.

The view from the Pentagon may, however, turn out to provide an accurate viewpoint. The current generation of armed UMA have been developed in the crucible of war. While the developers have been able to take advantage of new developments to extend their duration and increase the payload that can be carried, the current generation of armed UMA really do not need vast amounts of new investment.

What, however, is likely to happen is that their production lines may well be in operation for some time to come. The loss rate of UMA still poses problems for the designers. Replacements to maintain the operational footprint are therefore going to be an inevitable requirement but that is not the only reason why additional armed UMA are likely to continue being built.

With Al Qaeda showing all the signs of exploiting ungoverned places around the world for many years to come, the armed UMA will remain the weapon of choice to ensure governments can disrupt their activities at source. If the nightmare scenario of a follow-on attack to 11 September 2001 is to be prevented, some form of disruption of activities in terrorist training camps has to occur. To do away with armed UMA and to hope that terrorists will somehow not use additional freedom of manoeuvre that they would then enjoy to plan and conduct major attacks against the west seems a wholly naive view. Transnational terrorists have no track record of responding positively to any such gestures. Despite all the attempts in some sections of the media to taint the operations of armed UMA, they will remain, for the foreseeable future, 'the only game in town' when it comes to reaching into the remote sanctuaries in which international terrorists hide.

APPENDIX A

Analysis of UMA Strikes in Pakistan

In 2011 in the wake of the death of Osama bin Laden several high-ranking political leaders in the west started to write the epitaph of Al Qaeda. As far as they were concerned the organization had reached a tipping-point from which it could not return. Central to those claims was the impact that armed UMA strikes were having on the organization. Detailed analysis of such claims provides conflicting evidence as to their efficacy.

Of all the theatres in which armed UMA strikes occur, those in Pakistan fuel the most emotive reactions in the international media. Much of the coverage that ensues is ill-informed and of a generalist nature. Claims made in the media often solicit a muted response from White House officials. Off-the-record briefings provide indications of the success of the attacks, naming key players in various networks as having been successfully targeted. However, due to the lack of transparency over the nature of the strikes, presumably linked to an inherent nervousness over their legitimacy, it is the media that sets the agenda on the issue. Their inflated claims provoke a fierce and understandable emotional reaction.

One thing the media does little to address is the complex nature of the various groups involved in the insurgency. What is often referred to in general terms in the media as the Taliban is in fact composed of a series of groups, each with a slightly different agenda. Some, such as the Haqqani network, use Pakistan as a sanctuary from which to mount attacks in Afghanistan. To date they have not indicated a desire to use the sanctuary as a place from which to mount attacks on the international stage. Yet they have been subjected to eighty-eight (25 per cent) of the total number (352) of armed UMA strikes in Pakistan, according to the Long War Journal website, up until 29 November 2013. Fourteen of their leaders associated with the Taliban and Al Qaeda have been killed in this period out of a total of fifty-three reported by the Long War Journal.

Clearly the American agenda for attacking groups in Pakistan does not solely depend upon their interest in being involved in international terrorism. However, the focus of attacks upon the Haqqani network has been decreasing as a percentage of the total number of attacks. Up until July 2013 it had only been the target of a UMA strike on two occasions. That said, on 6 September 2013 seven members of the Haqqani network were killed when two missiles engaged a compound in the village of Dargah Mandi in North Waziristan, a known stronghold of the group. This is where a training facility known as Nawab Camp was located.

Pakistani officials were quick to condemn the attack but they also provided a statement on the strike saying that all of those who died were insurgents. One of those was subsequently believed, in reporting provided by the Long War Journal, to have been Mullah Sangeen Zadran, a Haqqani network commander who also served as the Taliban's shadow governor for the nearby Paktika Province in Afghanistan. Two mid-level Al Qaeda commanders from Jordan also died in the strike.

Some of the Taliban groups operating inside Pakistan do so under the shelter of the Pakistani intelligence service. They are widely regarded as being 'good Taliban'. Others, such as Lashkar-e-Taiba (LeT) – Soldiers of the Pure – are seen in a different light. Despite obvious connections to the Pakistani intelligence networks, they are disavowed. In January 2002 Pakistan's former president, Pervez Musharraf, banned LeT. This was in the wake of the attacks on 11 September in the United States. Despite the ban, the leadership of LeT still enjoyed some unofficial support from the Pakistani government. While restrictions were placed on their movements, the group continued to enjoy some manoeuvre room to plan attacks in India.

They are widely believed to have been behind the attacks in Mumbai in 2008 which saw 174 people die. Suspicion fell on the group due to their focus on the struggle over the sovereignty of Kashmir. For LeT the remote areas of Pakistan were a good place from which to plan operations against Indian forces in Kashmir. For the core remnants of Al Qaeda, the north-west of Pakistan provided an obvious place to move to when they had to leave Afghanistan. It was from there that they could orchestrate future attacks against the west through their increasing network of franchises.

Of all the groups under threat of armed UMA strikes it is the faction led by Hafiz Gul Bahadur that has attracted the greatest number of attacks (ninety-nine, or 28 per cent) from 2004 until 22 November 2013. The largest number of attacks against his group occurred in 2010 when fifty-three of the reported 117 attacks documented by the Long War Journal were directed at them. The rate of strikes against his group shows little sign of abating as seven of the eighteen attacks reported up until the end of July 2013 were targeted towards them. In 2012 that ratio had been fifteen out of a total of forty-six which was slightly up on the fourteen out of sixty-four in 2011.

Gul Bahadur gained notoriety in Afghanistan fighting against the Soviet armed forces during the occupation. As Pakistani ground forces encroached upon his territory he negotiated a peace treaty that required him to remove any foreign militants from South Waziristan. That was to lead him to align more closely with the Pakistani government. On several occasions he publicly encouraged Baitullah Mehsud to stop attacks upon the Pakistani forces fighting insurgents in North Waziristan. In 2008 the highly tenuous agreement with the Pakistani government came under strain as a result of armed UMA strikes.

In March 2011 he also threatened to withdraw from the agreement when one of his own commanders, Sherabat Khan Wazir, was killed in an armed UMA strike in Datta Khel. In February 2009 the 52-year-old was involved in the formation of the group Tehrik-i-Taliban (TTP). Previously he and Mullah Nazir had collaborated in July 2008 in the formation of Muqami Tehrik-e-Taliban (the Local Taliban Movement). When this group was established Gul Bahadur was named as its leader. He has also worked closely with Sirajuddin Haqqani. These shifting alliances are typical of the region and make analysis of the focal point of the armed UMA strikes difficult.

TTP is the group led by Baitullah Mehsud that is behind many of the large-scale acts of terrorism inside Pakistan including the assassination of Benazir Bhutto. Mehsud forged TTP by agreeing for it to amalgamate with a number of other groups that had similar objectives. One of those involved was Mullah Maulvi Nazir, the leader of the Ahmedzai Wazirs in South Waziristan. His group has been subjected to fifty-two (15 per cent) of the armed UMA strikes in Pakistan since 2004. The peak of activity against his group occurred in 2011 when nineteen of the total of sixty-four attacks were directed at them.

By contrast Mehsud's group had only been targeted on thirty-five occasions (10 per cent) over the same period. By attacking TTP the Americans were actually aiding their Pakistani colleagues. Mehsud was killed in an armed UMA strike on 5 August 2009 in the Zangar area of South Waziristan. The relatively low level of attacks directed at Mehsud suggests that the Americans were not that keen to help the Pakistanis combat the insurgency that threatened their own government's existence. The death of Baitullah Mehsud may also have been a factor as the rate of attacks dropped significantly post-2009. Increasingly the American focus was on those using Pakistan as a sanctuary for attacks against NATO forces operating in Afghanistan.

As United States forces were surged by President Obama into Afghanistan, so the tempo of armed UMA strikes in Pakistan initially grew. The linkage between the two clearly suggests that the political rhetoric pointing to attacks on Al Qaeda in Pakistan is somewhat superficial in nature. The armed UMA strikes in Pakistan seem more correlated with the objectives of the military campaign in Afghanistan.

Year	Number of UMA strikes
2004	1
2005	1
2006	3
2007	5
2008	35
2009	53
2010	117
2011	64
2012	46
2013	27

Table A.1: Number of UMA strikes in Pakistan [Source: Long War Journal]

Another key figure in the region is Abu Kasha al Iraqi. He is a go-between who links networks under the control of Al Qaeda with those of the Pakistani Taliban. The Long War Journal believes that sixty-one strikes have targeted his group over the period from 2004 to 29 November 2013. He is a key player in the area of Mir Ali in Pakistan. This is an area where many white Caucasian males who have been recruited to Al Qaeda base themselves. On 5 July 2011 an Australian national known as Saifullah was killed by two missiles fired from an MQ-9 Reaper. His death gave credence to claims that had been emerging at the time of a number of white jihadists operating in the area. This was the latest in a surge of attacks directed against white Caucasian males living in the Mir Ali area.

From 8 September 2010 to the attack that killed Saifullah it was reported that sixteen German males and two British men had been killed in armed UMA strikes. The two Britons were killed in Miranshah, another important geographic hub in the network of terrorist groups operating in Pakistan. Other men had also journeyed from France and other parts of Europe to become involved in Al Qaeda's activities. Americans have also been reported in the area and over 100 German nationals are believed to have travelled into Pakistan to become involved in terrorism planning. The aim would appear to be to equip these men with the skills for them to return to Europe and conduct an attack similar to that in Mumbai.

Clearly Al Qaeda saw benefits in recruiting white Caucasian males who could pass any stereotypical profiling activity. In another attack in the same area another Briton was also killed. He was known as Abdul Jabbar. He was reported as being referred to as the chief operational commander of the Islamic Army of Great Britain. Jabbar was a British citizen who had originated from the Jhelum district of Punjab. He had apparently survived a previous attack on 8 September 2010. Apparently Jabbar's younger sibling survived the attack that saw his brother killed.

Table A.1 shows the number of armed UMA strikes in Pakistan from 2004 up until the middle of July 2013. The rapid build-up of the use of armed UMA strikes in the wake of the election of President Obama is self-evident in the figures. What is also clear is the recent significant downturn in the rate of strikes that has taken place. However, the increase in armed UMA strikes did not happen as soon as President Obama moved into the White House. Indeed, during his first month in office only two attacks were noted in reporting produced at the time by the BBC.

A slight surge to five attacks in March and April was followed by another reduction in June 2009 when only three strikes were recorded. As the Pakistani military built up to the attacks on the insurgent strongholds in South Waziristan that were launched in the middle of September 2010 the armed UMA strikes tailed off from a new peak of six in September to two in October.

As the Pakistani army offensive unfolded in South Waziristan towards the end of the year President Obama authorized an increasing number of attacks in Pakistan. Despite the impact of the efforts made by the Pakistani army, eighty-three of the 352 attacks (24 per cent) have taken place in South Waziristan with specific peaks of activity in 2009 and 2011 either side of the moves made by Islamabad to create governance in the area.

The vast majority of these attacks (252 out of the 352 or 71 per cent of those recorded by the Long War Journal) have taken place in North Waziristan, an area of Pakistan where the central government is virtually unable and unwilling to create the levels of governance required to further reduce the manoeuvre room of the Taliban and their Al Qaeda associates. This creates a situation where UMA strikes are the only means by which the insurgents can be effectively targeted. Until the government of Pakistan is able to mount a coherent and long-lasting attempt to recreate governance over North Waziristan it is very unlikely, despite the reassuring political words emerging from Washington, that any significant downsizing of UMA strikes will occur.

Aside from a slight time delay, the increase in attacks authorized by President Obama in 2010 mirrored an increase in insurgent activity with attacks by Islamic militants rapidly increasing in reaction to the ground and air operations mounted by the Pakistanis. The point at which President Obama's increasing dependence upon armed UMA strikes peaked is self-evident from the figures. In January 2010 thirteen attacks were carried out in Pakistan. What was to follow provided the backdrop to the most intensive use of armed UMA that has yet been seen in history.

By March 2010 the inevitable backlash in the media had started with the Washington-based New America Foundation claiming in a report called 'The Year of the Drone' that 32 per cent of those who died in such attacks in Pakistan were civilians. Their analysis studied 114 attacks and concluded that more than 1,200 had died. However, subsequent reporting in July 2013 by the Bureau of Investigative Journalism provided a very different picture.

They noted that in 2010 a total of eighty-four civilians had died and nineteen children were also killed out of a total of 977 people who had died in armed UMA strikes. The civilian and child casualties therefore comprised 10.5 per cent of the total. While the total killed was not too dissimilar to the figures published by the New America Foundation, the ratio of civilians and children to insurgents was dramatically at variance with the Bureau of Investigative Journalism, reporting 874 insurgents killed.

The peak figure of civilian casualties (including children) occurred in 2009 when 136 died in armed UMA strikes out of an estimated total of 609. Even this figure was only 22 per cent of the total. The ratio of civilian and child deaths to armed UMA strikes in 2009 was 2.57. In 2010 it was 0.88. By contrast the reported death toll of insurgents to armed UMA strikes was 8.9 in 2009 and 7.47 in 2010. By 2011 the ratio of civilian deaths being reported by the Bureau of Investigative Journalism had fallen to fifty-eight in sixty-four strikes: a ratio of 0.9. The figures from 2012 where eight civilians died in a reported forty-six strikes produced a ratio of 0.17.

Up until 24 July 2013 no civilians have been reported as being killed in the twenty-seven armed UMA strikes carried out in the year. This was a trend that was unlikely to last. Subsequent reporting in the year to 5 September suggested that eleven civilians had been killed in UMA strikes. This is a ratio of 0.55; significantly up on the previous downward trend. In the same period a total of ninety-two members of the Taliban and Al Qaeda had also been killed in the twenty-seven strikes recorded in 2013.

On the basis of these numbers the assertions in the media that President Obama had somehow unleashed an indiscriminate killing spree upon the Pakistani population in the areas where the extremists sought sanctuary seems exaggerated. What happened in 2006 when the number of civilian (including children) deaths was reported to be 163 was clearly an exception, not the rule. Since President Obama entered the White House the pressure to reduce civilian casualties has clearly had an impact.

This is a point now conceded by one of President Obama's harshest critics, Chris Woods, who heads the covert war programme for the Bureau of Investigative Journalism. In comments reported in the *Guardian* in May 2013 Woods is quoted as saying: 'For those who are opposed to drone strikes there is historic merit to the charge of significant civilian deaths.' He went on to add: 'But from a contemporary standpoint the numbers just aren't there.' For one who has been such a vocal critic of the Obama administration, that is quite a statement.

The peak of 117 attacks in 2010 followed by the death of Osama bin Laden at his compound in Abbottabad in Pakistan on 2 May 2011 was the catalyst for the optimistic assertions emerging from Washington. This was not the first time the White House had issued such assessments. In March 2010 Leon Panetta, the director of the CIA, stated that: 'United States and Pakistani attacks against Al Qaeda in Pakistan have crippled the group and its leadership.'

While it is clear from subsequent analysis that Al Qaeda's activities were being severely disrupted, to suggest that their leadership was crippled was political hubris. Up until the end of 2010 a total of twenty-six attacks had been planned and disrupted by United States officials in conjunction with their overseas colleagues since 11 September 2001. In the next two years Al Qaeda attempted the same number of attacks. The increased tempo of Al Qaeda activity showed that Panetta's remarks were perhaps a little too optimistic.

The leadership of Al Qaeda was relatively quick to respond to the death of its leader in May 2011. The organization was clearly prepared for such an event. Taking its lead from business in the same way that it has managed to brand itself over the past decade, the group appointed Dr Ayman al-Zawahiri as the new emir of Al Qaeda. Selling that idea to the complex tapestry of franchises that had sprung up under Osama bin Laden took a little longer. Some of the groups appeared reluctant to endorse the new leader. Several cited the oath of allegiance to Bin Laden that they had provided. This was to the man, not necessarily the organization. Dr Zawahiri's often authoritarian approach to his views on Islam did not endear him to the wider base of supporters. He was often seen jabbing his finger at the screen on many of the videos where he tried to communicate his views to his audience.

The rate at which the armed UMA strikes started to take their toll on the leadership of Al Qaeda became a pressing problem for Dr Zawahiri. He ordered those who had survived the wave of attacks in 2010 to lie low for a period of time. Second-tier members of the core group of Al Qaeda residing in Pakistan were promoted to new positions and new volunteers were recruited to replace those who had been killed and those who had stepped into new positions. It was inevitable that it would take time for them to become effective members of the organization.

While Dr Zawahiri and his new leadership team laid low, it might reasonably be assumed that the threat to the west had reduced, albeit temporarily. In fact, that was far from the case. Al Qaeda had already been operating a new strategy for some time. Instead of asking people to travel overseas to attend terrorist training camps, they issued notices through various internet forums and publications suggesting that would-be jihadists should stay at home and conduct random acts of violence. The killing of Drummer Lee Rigby in Woolwich in 2013 is one example of the perpetrators responding to such calls.

Not all would-be jihadists have obeyed the directive. As recent court cases in the United Kingdom reveal, a number of groups have still managed to send key players overseas to establish links with people linked to Al Qaeda. This follows the route taken by two of the suicide bombers involved in the attacks on London in 2005. However, the impact of the armed UMA strikes is clear in the evidence that has emerged. Several of those who travelled to Pakistan have noted the austere conditions under which they lived. Fear of strikes by armed UMA has clearly had an effect upon the operational manoeuvre room enjoyed by Al Qaeda and its cohorts in Pakistan.

Nevertheless, some still managed to visit the training facilities and return to the United Kingdom prepared to carry out acts of extreme violence. In one case reported in Birmingham the terrorists had planned to detonate up to ten devices in a short period of time across the West Midlands. This kind of time-compression tactic has usually been seen at work in Iraq where multiple almost simultaneous attacks occur to stretch the responses of the emergency services.

Al Qaeda now operates a multi-tiered organization. Part of that is based upon trusted networks. Some of the key franchises are based on leaders with long-standing links to what is known as Al Qaeda's core leadership in Pakistan. The remainder is formed of a series of people and groups with loose affiliation to the overall ideology expressed by the group's leadership. At one end of the spectrum occupied by these loose affiliates lies the individual or self-motivated lone wolf. The alternative to this is a group of people who have come together to become involved in terrorism. There is little doubt that this evolution of Al Qaeda as an organization can be directly attributed to the impact of the armed UMA strikes in Pakistan, the Yemen and Somalia.

But what of the evidence that the armed UMA strikes create a backlash in local communities, radicalizing people who join up with militant groups? Analysis of data derived from the *IHS-Jane's* Joint Terrorism Intelligence Centre (JTIC) of terrorist activity in Pakistan provides an inconclusive picture.

Year	UMA Strikes	Estimated Death Toll	Civilian Death Toll	Ratio Death/ Strikes	Damaging Attacks
2008	35	252	81	7.2	1216
2009	53	473	136	8.92	1405
2010	117	874	103	7.47	1263
2011	64	447	58	6.98	2150
2012	46	238	8	5.17	2156
2013	27	74	0	4.35	1920

Table A.2: Comparison of armed UMA strikes and damaging attacks in Pakistan

The term used by *IHS-Jane's* to describe acts of violence is 'damaging attacks'. They embrace all forms of physical violence from kidnapping, throwing grenades into crowded rooms or areas, drive-by shootings, assassinations, the use of rockets or mortars and multiple forms of suicide and IED attacks.

If people in Pakistan were actually being radicalized by armed UMA strikes it might be possible to see some of the rhetoric expressed as outpourings of grief or anger on the streets in the wake of an attack converted into increased levels of insurgent activity. From a detailed month-by-month analysis of the overall levels of attacks there appears to be little correlation between the two datasets. On investigation the highest degree of correlation occurs when the two datasets are offset by ten months. The figure of just over 0.5 that is produced is hardly convincing. Why would there be a ten-month delay in reacting to armed UMA strikes?

Tables comparing the number of armed UMA strikes with the level of damaging attacks carried out by insurgent groups on a month-by-month basis in Pakistan produces little in the way of visual clues as to any direct correlation. Indeed, as the armed UMA strikes peak in the latter part of 2010 there is a notable reduction in the levels of damaging attacks over the next year. Equally, as the armed UMA strikes taper off towards the end of the reporting period there is no equivalent reduction in the level of insurgent activity in Pakistan.

When the ratio of damaging attacks to armed UMA strikes is plotted, the coercive effect of the attacks also becomes clear. As UMA strikes peaked in 2010 there was a lasting and perceptible decline in the ratio of damaging attacks to the number of attacks conducted by UMA. This pattern did not start to reverse significantly until the UMA strikes started to fall away in 2012 and 2013. At this point the ratio measure began to increase again above the levels seen in 2009.

Indeed, when the average number of damaging attacks per day is used as a measure the impact of UMA strikes again is visible as the average goes from a figure of around five per day in 2008 and 2009 to below three per day in 2010 and 2011. As the UMA strikes taper off from 2011 the average number of damaging attacks per day returns to an average of around four. Up to September 2013 it has not managed to recover to the early levels recorded in 2008 and 2009, suggesting there has been some long-term decline in terrorist activity in Pakistan. To attribute all of that to UMA strikes would be foolish but these figures do not provide any succour to those who suggest that UMA strikes were likely to increase the levels of violence. This again supports the separate analysis carried out by the RAND Corporation team.

One important factor that must not be overlooked in Pakistan is the degree to which the damaging attacks that occur reflect local issues. The insurgent landscape in Pakistan is a complex pastiche of historical and other contemporary insurgencies. To really try to answer the question as to whether armed UMA strikes are having a notable impact in encouraging more people into the insurgency, it is important to break down the levels of damaging attacks by geography.

APPENDIX A

The level of damaging attacks that have been recorded in the FATA bears little resemblance to the pattern of armed UMA strikes. Of all the areas of Pakistan where a signature showing a positive correlation between the two should emerge, the FATA would be one of the two areas high on the list. The other is the NWFP. There too the figures are not clear. Indeed, the levels of damaging attacks in 2008 and 2009 are dramatically reduced in 2010, only starting to recover slightly in 2011. Yet again there is some suggestion in the figures that the coercive effect of the strikes has deterred insurgent activity. What the analysis of the figures does prove conclusively is that there is no appreciable upturn in violence in those areas of Pakistan where UMA strikes occur.

There is also little to suggest that some form of displacement effect has occurred with Taliban or Al Qaeda followers moving into new geographic areas to exact revenge for the UMA strikes. Correlation analysis conducted between the pattern of UMA strikes over the period with the frequency of damaging attacks in each of the agencies of the FATA shows little coupling between the two variables.

The highest correlation occurs with damaging attacks in the Mohmand Agency (value 0.35). The remaining figures are all below 0.15 and therefore are not statistically significant. This is especially important for the analysis of the correlation in North and South Waziristan where the majority of UMA strikes occur and some form of local backlash might have been expected. Indeed, there is evidence in the analysis to suggest that after a relatively quiet period where few UMA strikes occurred towards the end of 2009, damaging attacks increased markedly after a short delay. Once the UMA strikes ramped up in 2010 the level of terrorism activity, particularly in South Waziristan, dropped to virtually zero for a period of several months; further evidence of the coercive effect of the UMA strikes.

This conclusion is borne out by work published by the RAND Corporation in July 2013. They draw similar conclusions on the basis of a very detailed empirical study of the relationship between UMA strikes and reported insurgent/terrorist activity on the ground in Pakistan.

The situation in Balochistan is somewhat different. Appendix C explores the situation in that province in greater detail as the analysis examines the degree to which attacks on NATO tankers shipping fuel to Afghanistan are an indicator of the insurgency enacting retribution for the armed UMA strikes. Visual inspection also reveals little in the way of any correlation between the levels of damaging attacks and armed UMA strikes. In Karachi, where sectarian violence dominates the security landscape, there is also little evidence to suggest that armed UMA strikes are having a radicalizing effect.

One comment often made by those concerned about the impact of armed UMA strikes is that the death of civilians is a particular radicalizing element. Certainly the protests on the streets of Pakistan after some of the attacks that have claimed the most lives provide visible proof of the degree of anguish being experienced. But does that turn to a desire for revenge? What people say is one thing. What they actually do when confronted by choices is sometimes quite different.

The International Bureau of Investigative Journalism has gone to great lengths to try to estimate the death tolls arising from armed UMA strikes. They are the source of the figures produced in Table A.2. Understandably, the picture mirrors that of the armed UMA strikes. It is entirely reasonable to suggest that the more strikes occur, the greater the casualties. However, a detailed analysis of the trends of the ratio of estimated deaths to armed UMA strikes shows an interesting decline. It is possible to suggest that this is the result of a greater degree of control being brought to the decision-making processes since the Obama administration entered the White House.

The downward trend is also perhaps an indication of a more careful approach being adopted towards 'signature strikes'. These carry the greatest risk of civilian casualties. While it is hard to obtain specific details of the numbers of 'signature attacks' that took place in 2010, it seems reasonable to suggest that some of the 117 attacks that occurred in that year were of that form. However, the detailed break-down produced by the Bureau of Investigative Journalism provides a counterweight to that argument.

Year	North Waziristan	South Waziristan	Other Areas	Totals
2004	0	1	0	1
2005	1	0	0	1
2006	1	0	2	3
2007	4	1	0	5
2008	18	14	3	35
2009	22	27	4	53
2010	104	7	6	117
2011	41	22	1	64
2012	39	6	1	46
2013	15	5	0	20
Totals	**245**	**83**	**17**	**345**

Table A.3: Geographic distribution of armed UMA strikes in Pakistan by year [Source: Long War Journal]

APPENDIX A

The civilian death tolls, which include reports of children that have been killed, peaked in 2009. In 2010, when the Obama administration has been accused of adopting what might be called a much more liberal approach to the use of armed UMA strikes, the level of civilian casualties estimated by the International Bureau of Investigative Journalism actually slightly decreased. These figures are based upon the upper estimates of civilian casualties produced by the Bureau. Their methodical approach has also produced minimum estimates that are notably lower for the period from 2008 to 2011 and very marginally lower in 2012–2013.

This may offer an explanation as to why the ratios have subsequently reduced as the pilots and weapon systems operators using the UMA have become better at recognizing the kind of behavioural indicators that suggest terrorist activity. Harder targeting rules brought in by the Obama administration, in part to try to address critics of the strikes, are also demonstrably having an effect.

One other important factor to consider when looking at the armed UMA strikes in Pakistan is their geographic distribution. Table A.3 shows that the main focus of the attacks has been in North Waziristan where 71 per cent of the strikes have taken place since their inception. This remains the location in Pakistan at the epicentre of armed UMA strikes. The pattern of attacks does show that the focal point is narrowing.

In his first visit to Pakistan at the beginning of August 2013 since becoming the United States Secretary of State, John Kerry made a number of remarks designed to placate critics of the attacks. While refusing to suggest a date when the tactic would no longer be required, he did suggest that the number of targets in the area was diminishing. It would seem that in the short term the trend of armed UMA strikes will remain slightly downward or even flat.

To suggest that none will ever be needed again would be to fly in the face of logic. If the threat is removed, Al Qaeda and its associates will simply regroup in North Waziristan. Given that all the indicators are that the remaining senior leadership of Al Qaeda still regards North Waziristan as the best place for them to hide, it would be foolish in the extreme, given the past history, to suggest that they are not capable of staging a comeback.

Analysis of UMA Strikes
in the Yemen

While the headlines about so-called 'drone' strikes often focus on what is happening in Pakistan, it is increasingly likely that in the coming months the activities of armed UMA over Pakistan will continue to decrease. In the Yemen, however, the trend may not mirror that in Pakistan. On 20 January 2013 two strikes thought to have been conducted by armed UMA killed thirteen militants. Ten died at what was reported to be a bomb-making factory in the province of al-Bayda. Three others died in an attack in central Marib Province against a car that was thought to be carrying extremists. The initial tempo in both Pakistan and the Yemen signalled another busy year for armed UMA strikes in both countries.

One illustration of that emerges vividly from the heightened terrorist alert levels that arose at the start of August 2013. Reporting suggested that a teleconference held between the core leadership of Al Qaeda and the main leaders of the franchises had resulted in specific intelligence being obtained of a threat to American embassy facilities across the Middle East. The appointment of the leader of Al Qaeda in the Arabian Peninsula (AQAP) as the overall second-in-command of Al Qaeda was also thought to be a trigger for an increased level of caution vis-à-vis embassy staff, many of whom were swiftly evacuated. Other countries quickly followed suit. Apparently other intelligence sources had reported a surge of arrivals of people linked to Al Qaeda into the Yemen towards the end of July. Clearly something was being hatched.

The closure of the embassies was heralded as a precautious approach in the wake of the attack by Al Qaeda elements on the consulate in Benghazi in Libya in which the US ambassador was killed. Within days of the threat level being increased, reporting emerged of an upsurge in armed UMA strikes. Over a twelve-day period eight armed UMA strikes took place in the Yemen. Evoking an image from the First World War, this spate of attacks was labelled as a barrage by some sections of the media. Such hyperbole is ill-advised and hardly objective. Those soldiers who survived the First World War would not have thought eight strikes by missiles fired from a UMA could be classified as a barrage.

On 27 July three UMA strikes occurred, coinciding with the Eid Festival that marks the end of Ramadan. These attacks killed a reported thirty-four suspected militants. The first of the three attacks occurred in the early hours at a place called Wadi Abida in the central province of Marib. Reports suggested that six Al Qaeda-affiliated militants died in the raid. This was the latest in a series of strikes in what are very remote oasis-fed farms in the area surrounded by the desert. The second and third attacks took place in the eastern region of the Hadramaut.

These attacks came in the wake of an announcement by the Yemeni government that they had foiled an attack by Al Qaeda-affiliated militants to seize the port of Al Mukalla. This is the fifth-largest city in Yemen and home to two major oil and gas export terminals on which the economy of the country is hugely dependent. This appears to have been an attempt to repeat the attack on the In Amenas gas facility in Algeria. This saw thirty-seven people lose their lives when local Special Forces intervened as hostages were being transported from the accommodation blocks inside the complex. It seems that the intent of the terrorists was to destroy the facility and kill all the hostages. On 30 July four more Al Qaeda-linked fighters were killed when the car in which they were travelling was destroyed by a UMA strike in Marib Province. All those who died were from the Yemen.

Previous patterns of armed UMA strikes have also surged in response to either specific intelligence or as a result of AQAP showing its intent to attack international targets. By 8 August 2013 the Bureau of Investigative Journalism had noted twenty-seven armed UMA strikes in the Yemen during the year. Eight of these occurred in the Marib Province with five in Abyan. Units affiliated to AQAP operating in the Hadramaut received three attacks in a narrow window at the start of August 2013.

On 6 August one of a number of leading AQAP figures thought to be involved in planning a major attack in the Yemen, Saleh al-Tays al-Waeli, was killed in an armed UMA strike in Marib Province. His was one of twenty-five names on a list produced by senior Yemeni officials as a result of the increased alert levels. This was followed by a series of armed UMA strikes at several locations across the Yemen aimed at further disrupting the activities of AQAP and its top leadership. A strike on 11 August was the tenth since the president of the Yemen visited Washington on 1 August 2013. This brought the total for the year to twenty-two armed strikes in the Yemen, placing it clearly ahead of the eighteen in Pakistan over the same period.

The situation in the two theatres, however, does have some important differences. Al Qaeda enjoys a greater degree of manoeuvre room in the Yemen. The area in which they now operate is approximately ten times the size of the current Al Qaeda footprint in Pakistan. Analysis of the locations where armed UMA strikes are reported shows that they cover a large part of the Yemen. Another problem for the Americans is that the intelligence infrastructure is also less developed, making it harder to target key individuals in the leadership of the local franchise. That said, there have been some important success stories.

The death of Said al-Shihri (whose real name was Abu Sufyan al-Azdi), the deputy leader of Al Qaeda's franchise operating in the Yemen known as AQAP, is a case in point. He had just passed his fortieth birthday. His death was formally announced on 17 July 2013. Rumours had been circulating since the end of January that he had been killed, so the confirmation was not a surprise. It is the latest in a series of high-profile casualties to arise from strikes inside the Yemen.

At the end of July 2013 a total of twenty-one attacks had occurred in the Yemen according to reporting by the Bureau of Investigative Journalism and the Long War Journal. On a pro-rata basis that would provide an estimate of thirty-six attacks if the prevailing strike rate were to continue. This is not far behind the total for 2012 and shows little sign of tapering off. At this point the Yemen surpassed Pakistan as the primary centre of armed UMA strikes outside Afghanistan. Since the first armed UMA strike occurred in the Yemen in 2002, the latest round of attacks brought the overall total to eighty in which it is suggested that 384 enemy combatants had been killed alongside eighty-four civilians. If current trends in Pakistan and the Yemen continue, UMA strikes in both areas will not be vastly different, showing how the Obama administration is slowly changing its focus from Pakistan.

This is a point borne out when examining disclosures in *The Times* in 2011 on the location of airbases from which armed UMA now operate. According to the reporting in *The Times*, several potential bases have been established that create a ring of fire-power around the Yemen. The most easterly is the Seeb Air Base in Oman. In the United Arab Emirates the airbase at Liwa is also claimed to be a home for armed UMA. To the west in Djibouti the American base at Camp Lemonnier is also thought to be involved.

In addition to these the reporting in *The Times* also noted bases in Saudi

Arabia as being potential hosts for armed UMA. The radius of action from this set of bases provides excellent coverage over the Yemen and a significant percentage of Somalia. While it is possible to attach some credence to these claims other sources have gone further, pointing to potential operations being mounted from the southern city of Arba Minch in Ethiopia. The lack of transparency on armed UMA deployments makes such claims difficult to verify. However, from a purely geographic viewpoint the location provides excellent coverage over war-torn Somalia.

It would also allow the Americans to operate UMA in the area when Kenyan military forces are operating in the south-west of Somalia and help Ethiopian forces who are active on the ground against the Al Qaeda franchise Al-Shabab. The level of armed UMA strikes in Somalia has remained very low with some sources suggesting that between three to nine attacks in total have taken place. This is significantly below the total number of attacks in Pakistan and the Yemen. Unlike the Yemen, Somalia is not a focal point for the United States armed UMA strikes at present. The dynamic that could change that would be if any evidence started to emerge of the flight of senior Al Qaeda leaders into Somalia from the Yemen. Given the people-smuggling routes that exist between Somalia and the Yemen, that is not a prospect that can be dismissed out of hand.

Al-Shihri had been originally captured in Afghanistan in 2002 and detained in Guantanamo Bay. After his release into the custody of the Saudi authorities who had earmarked him to attend their de-radicalization programme he fled to Iran before appearing in the Yemen. He was named as the deputy to the leader of AQAP Nasir al-Wuhayshi in February 2009 when the franchises operating in Saudi Arabia and the Yemen merged. He and three others appeared in a video posted on the internet at the time announcing the formation of AQAP. Several other high-profile individuals released from the Cuban detention facility have made the same journey to join AQAP. The membership of the franchise was also helped when many of its supporters were able to break out of a Yemeni prison in 2006.

Al-Shihri had previously been unsuccessfully targeted on several occasions. In October 2012 he appeared in another internet video to show that claims of his death in an air-strike had been premature. His bravado, however, was short-lived. His death sends a message to the leadership of the local franchise. The Americans are well aware of their local, regional and international objectives to conduct acts of terrorism. The franchise will therefore remain a high priority as far as the allocation of United States counter-terrorism resources is concerned.

One specific difference between the operations in Pakistan and the Yemen is the degree to which the United States military takes responsibility for targeting decisions. Up until early 2013 operations in Pakistan were run by the CIA. In the Yemen the operations were controlled and executed by the United States military. For many concerned at the way in which the CIA had moved away from being purely a spy agency and was getting involved in decisions to assassinate key Al Qaeda leaders, the model for command used in the Yemen was preferable.

From the viewpoint of terrorists Pakistan and the Yemen have one important thing in common. Both have vast areas that are remote from the capital city. Access to these areas is difficult, even for military forces. Law enforcement on the ground is provided by local people in the form of adhoc militias. Loyalty is to local tribal leaders. Central government provides little in the way of even basic services at the local level. These are areas where it is easier for Al Qaeda to establish training bases. The model of what was achieved in Afghanistan prior to 11 September still applies. In the Yemen the Hadramaut, the ancestral home of the Bin Laden family, is one such vast area of the eastern side of the country.

The Hadramaut occupies an area of nearly 200,000 square kilometres (75,000 square miles). Its population density of 10.5 per square kilometre provides the perfect backdrop against which to create a footprint on the ground. It is in these largely unpopulated areas of the Yemen under the protection of local tribal chiefs that Al Qaeda groups first started to establish their franchise. Given its vast uninhabited area the Hadramaut provides an excellent backdrop for AQAP and other members of Al Qaeda to try to establish a footprint. From the middle of May 2012 the United States launched seven attacks in the area until the end of that year. What then followed was a period of no armed UMA activity until August 2013. During this period other provinces in the Yemen were in the cross wires, such as Shabwah, Al Jawf and Al-Bayda.

Some key players in the local franchise had previously been active in Saudi Arabia. As their campaign in the kingdom came under sustained pressure from the Saudi authorities, they had to retreat over the border and attempt to re-group. There is little doubt that if they are given the time and space they will try to return across the border and conduct attacks against the oil infrastructure and government. In such a remote area it is difficult to put even Special Forces on the ground. Strikes by armed UMA are therefore an obvious tactic to disrupt the activities of the Al Qaeda-affiliated groups operating in the province. In 2012 seven of the forty-two armed UMA strikes that took place in the Yemen occurred in the province. Before May 2012 the area had remained free of UMA strikes.

Given the uncertain security situation in the Yemen it was only a matter of time before several areas of the country would come under the influence of Al Qaeda-linked groups. One indication of the freedom of manoeuvre enjoyed by Al Qaeda in the Yemen is the map of armed UMA strikes carried out over the period from 2009 to the present day.

There are only two areas of the country where armed UMA strikes have not taken place. One of these is the inhospitable east of the country along the border with Oman. The other is the far west along the Yemeni coastline of the Red Sea. The vast majority of armed UMA attacks have taken place in a central belt from Aden in the south to the Marib, with Abyan the area most in focus.

Once the terrorists established themselves in some of the more remote areas of the Yemen, it was almost inevitable that armed UMA strikes would feature as a component of an overall counter-terrorism response from the Yemeni government.

Year	Number of UMA Attacks
2002	1
2009	2
2010	4
2011	10
2012	42
2013	22

Table B.1: Number of UMA strikes in the Yemen [Source: Long War Journal]

Table B.1 shows the number of armed UMA strikes in the Yemen over the period immediately after the terrorist attacks on 11 September 2001 through to 11 August 2013. While armed UMA strikes may not quickly return to the level achieved in 2012, it seems unlikely that in the near future they will return to the levels prior to 2010.

In 2002 there was an isolated attack on a key member of Al Qaeda travelling in a convoy in the country. This specific attack by a Predator drone on one of the people thought to have been involved in the suicide attack on the USS *Cole* is often highlighted as the beginning of the use of armed drones to target members of terrorist groups. The victim of the attack was Qaed

Salim Sinan al-Harethi. He was regarded as a man close to the leadership of Al Qaeda and had reportedly spent some time with Osama bin Laden in the Sudan in the early 1990s. His car was travelling in a convoy and was struck by a Hellfire missile. Five other passengers in the vehicle were also killed. One of those was an American, whose name was Ahmed Hijazi. He was the first United States citizen to die in an armed UMA strike. History already shows that he will not be the last.

The attack took place in the northern province of Marib which is located around 100 miles to the east of the capital Sana'a. Outwardly its bucolic calm belies the reputation the province has for violence. Similar problems with rights over ownership of land exist here as well as in places like Afghanistan. The Marib is a place where Al Qaeda has managed to establish a durable footprint. This is somewhat surprising, given that this area is the location where the Yemen's limited oil and gas reserves are located. Yet locals complain that the majority of any income does not flow through to the population, instead remaining in the coffers of local government officials. This generates resentment at the Yemeni government.

Local people dislike the reputation that the Marib has gained for sheltering Al Qaeda. They point to areas where local tribal leaders have been unable to enforce a form of local governance which has created a security vacuum eagerly filled by Al Qaeda. Local people are then simply too weak to overcome an organized group, even when its resources are meagre. Re-establishing security in such regions is not easy. It is not simply a question of sending in the Yemeni army. This can lead to areas becoming safe havens for Al Qaeda that are then very difficult to dislodge.

Armed UMA attacks on potential terrorist facilities avoid the contentious issue of placing boots on the ground. However, it carries with it the danger that when mistakes are made and civilians are killed local people can become recruited to Al Qaeda's cause. One event that had this effect was an air-strike that killed Marib's deputy governor in May 2010. He had been trying to mediate with local tribal leaders to ensure they would not take up arms with Al Qaeda elements in the local area when he was killed. The attack provoked a furious local reaction, with pipelines and electricity pylons being destroyed.

Up until 8 August 2013 eight armed UMA strikes have been reported in the Marib. Prior to that point only four had taken place. Yemeni officials noted at the time that reports had emerged after the attack on the USS *Cole* that al-Harethi had taken refuge with a fellow conspirator Mohammed Hamdi al-Ahdal in a village called Hosun al-Jalal in the province of Marib. Local tribal leaders offered al-Harethi sanctuary from the Yemeni authorities. Local tribal leaders provided that sanctuary; not for ideological reasons but simply because they paid a rent of $30 a head for the mud house they were given.

By all accounts they lived quite separate lives, praying alone and spending most of their time on their computers and satellite phones. When Yemeni government officials arrived in the village asking questions as to the men's whereabouts they quietly departed in the middle of the night. The attack on the USS *Cole* had a devastating impact upon the Yemeni tourist industry with $1.5 billion in revenues being lost virtually overnight. For a country that can barely sustain its population, this was a heavy price to pay for the activities of AQAP.

In an interview with the broadcaster CNN the then Deputy Secretary of Defense Paul Wolfowitz noted that 'we've got to deny sanctuaries everywhere we're able to'. Today his words take even greater meaning as Al Qaeda franchises fan out across North Africa from their original bases in the Yemen and Somalia. Where instability creates ungoverned spaces in countries with a significant Muslim population, Al Qaeda will inevitably try to establish a presence on the ground. In Niger, Mauritania, Mali, North-West Nigeria, Western Sahara and Libya activities by groups linked to Al Qaeda have been on the increase.

The franchises in these countries are often created by local people. They are then joined by people who travel from nearby Arab states in search of jihad or from countries in Europe. To support the franchises in their early days, experienced Al Qaeda operatives also will travel hundreds of miles from across the Middle East and also other locations where the organization has an established footprint, such as the Caucasus.

For the next six years no UMA strikes were reported in the Yemen. In 2009 and 2010 a small number of attacks targeted key members of AQAP. That was the point at which the attacks were about to increase significantly. The main focal point for these was the Province of Abyan in the south where AQAP was fighting to establish a significant toehold on the ground.

The level of attacks in 2011 and 2012 are a reflection of three key factors. The first was the success that the United States had achieved in eliminating high-ranking Al Qaeda and Taliban leaders in Pakistan. If armed UMA strikes could achieve an impact in Pakistan, then it was obvious that same model could be transferred to the Yemen. However, doing that would take time. American officials signalled in August 2010 that armed UMA strikes would be increasing in the Yemen but another simple message was also sent out at the same time. Wherever Al Qaeda tried to find a sanctuary, the United States armed UMA strikes would follow; there would be no let-up in their pursuit of the terrorists. Today the consequences of that approach are now being felt by Al Qaeda franchises on the other side of Africa.

To conduct armed UMA strikes in the Yemen an extensive intelligence infrastructure needed to be established. That was necessary to avoid the backlash that would otherwise arise if large numbers of civilian casualties were to occur. The reaction to the death of the deputy governor in Marib Province had provided a clear indication of the smouldering resentment that existed just below the surface. HUMINT would inevitably be in short supply. In Pakistan the Americans had been able to build on long-standing ties and facilities. In the Yemen much of that had to be created. Developing HUMINT sources takes time. The ramp-up in attacks since 2011 reflects the increasing capabilities of the intelligence infrastructure.

The second was a growing concern in the United States about potential attacks on America that were sourced to the Yemeni-based franchise of Al Qaeda called Al Qaeda in the Arabian Peninsula (AQAP). In 2008 two suicide bombers had blown themselves up outside the United States embassy in the capital Sana'a. Sixteen people died in that attack. In 2009 four South Korean tourists were killed and AQAP quickly claimed responsibility. An attempted assassination of the British ambassador in Sana'a in April 2010 was also linked to AQAP.

The local attacks in the Yemen were a presage to what was to follow as AQAP embarked upon a series of audacious attempts to project its image onto the international stage. The first of those involved a young Nigerian man who had been trained in the Yemen trying to blow up an airliner as it descended into Detroit on Christmas Day 2009. The individual involved, Umar Farouk Abdulmutallab, was a 23-year-old man who was the son of a wealthy Nigerian banker. His journey into becoming radicalized had involved attending a degree course at University College London where he graduated with a degree in mechanical engineering in June 2008.

During his time in London he had become the president of the university's Islamic Society. While appearing briefly on the radar horizon of the British Security Service he had left London to journey via Nigeria to the Yemen in 2009. There he became involved with AQAP and was selected to be the person to carry what was known as the 'underpants bomb' onto the plane bound for America. Due to a series of intelligence failures he was not prevented from attempting to carry out his mission.

Days before he attempted to blow up the flight from Amsterdam into Detroit, two training camps in Sana'a and Abyan Provinces in the Yemen had been targeted by cruise missiles fired from a United States submarine. The attack was coordinated with Yemeni security forces and saw thirty-four members of AQAP killed and seventeen arrested. The leader of AQAP in

Abyan Province, Muhammad al-Salih al-Awlaqi, was also reported to have been killed alongside two other local commanders. AQAP immediately claimed that over sixty civilians had been killed in the same attacks. While those air strikes removed some key leaders, it did not deter AQAP from returning to the international arena and attempting another spectacular.

AQAP also admitted trying to send two parcel bombs to America a year later in what has become known as the 'Cargo Planes Bomb Plot'. Both of these attacks were seen to be innovative. The attempt to kill the Saudi Security Minister Prince Mohammed bin Nayef in August 2009 also came close to being successful. It was the first time the Saudi royal family had come into the cross wires of Al Qaeda. One common factor was shared by these three attacks. They tried to exploit loopholes in existing security arrangements and highlighted the continuing threat from AQAP.

These devices were built by Ibrahim al-Asiri. He is Al Qaeda's leading bomb-maker. To date he has survived at least two armed UMA strikes. Innovation is his watchword, with devices being built to evade existing airport security detectors. After the failure of the attack in 2009 that was aimed at Detroit, Al-Asiri went back to the drawing board and manufactured a new variant of the so-called 'underpants bomb'.

The latest version, which was intercepted after an intelligence source revealed its existence, had 300 grams of pentaerythritol tetranitrate (PETN). The original device only used 80 grams of PETN. This makes the latest version even more deadly. It is the same amount of PETN used in the 'Cargo Planes Bomb Plot' which even highly-trained sniffer dogs were unable to detect. The substance is so stable that few of its molecules 'leak' to the outside air to create a signature that can be detected.

Al-Asiri has proved himself to be a highly-accomplished innovator when it comes to designing and building bombs. Having trained as a chemist, he is a professional in his field. It is therefore of little surprise that he is highly placed on the targeting list for armed UMA. On at least two occasions the CIA thought they had managed to kill him but he managed to evade the attacks. Since the deputy leader of AQAP was killed in another attack due to lax security measures when using a mobile phone, Al-Asiri has gone to ground.

The major concern is that he is passing on his skills to others, who may then use those to continue the AQAP international campaign irrespective of whether Al-Asiri manages to avoid being killed. To date AQAP has proven to be a robust organization that has still continued to function even when several of its high-profile leaders have been killed.

In one of their more obvious attempts to send out a message of defiance, AQAP conducted an attack against the Yemeni intelligence authorities in June 2010. Eleven people, including a child, died in the attack which saw some of its members freed from captivity. The attack happened just after dawn and involved four men carrying machine guns and rocket-propelled grenades.

For the United States any developments in the Yemen that saw AQAP gain a greater foothold in the country were an obvious source of anxiety. This would allow Al-Asiri to have greater freedom of manoeuvre. It would also allow him to train other potential bomb-makers. While some Al Qaeda franchises clearly focused on domestic activities, AQAP made their international and regional intentions clear.

With the origins of the group partly anchored in Saudi Arabia it was also likely that any increased Al Qaeda presence in the Yemen would have also worried the leadership in Jeddah. In one specifically intense period at the start of June 2011 the United States was undertaking armed UMA attacks in the Yemen at a rate of nearly one a day. Abyan Province was an area that received at least six attacks in that period.

The third factor was driven by the levels of instability in the Yemen and the impact on the country of the Arab Spring. That provided a huge opportunity for AQAP to try to consolidate its hold on various areas in the south of the country. A brief attempt to establish a new state-within-a-state in southern Yemen was eventually overcome by the Yemeni security forces.

Recapturing the land occupied by AQAP took time as elite units of the Yemeni armed forces had to divide their time between putting down the spontaneous uprising against President Saleh that occurred as a result of the Arab Spring and other long-standing sub-regional grievances existing between tribes that occupy the north and south of the country. The country also suffers from the effects of a demographic bulge with many of its citizens being below 35 years of age.

The lack of employment and the inadequacies of the Saleh regime provided some of the catalytic points behind the protests that engulfed the country in 2011. Add to this toxic mix the Yemen's struggle to find sufficient ground water to supply its growing population and the reasons why the Yemeni government was grateful for United States intervention in the form of armed UMA strikes is understandable. With the Yemeni government preoccupied, the Americans had to step up the use of armed UMA in order to place pressure on AQAP. Public statements from the United States indicating that the Yemeni government have sanctioned the strikes against

AQAP and perhaps even provided vital intelligence information are denied by government officials in the capital Sana'a.

While the figures provided in the table above are sourced from the reputable Long War Journal, problems do exist in being certain about the exact nature of the attacks. Some may be falsely attributed to armed UMA. The United States has been known to target some terrorist training camps in the Yemen and elsewhere with cruise missiles fired from submarines. These are largely covert operations. It may therefore suit the White House for the numbers to be incorrectly attributed by non-governmental organizations.

Air strikes also occur over the areas where AQAP are active. These are conducted by the Yemeni Air Force. This shows the Yemeni government is also able to play its role in trying to ensure terrorist training camps do not gain a significant foothold in its country. The fear that they will replace the camps in Pakistan is very real. Migrations of key Al Qaeda leaders from Pakistan into the Yemen have also been noted.

After the failure of Al Qaeda to exploit the security lacuna in the Yemen brought about by the uprisings associated with the Arab Spring, the group has inevitably turned to an asymmetric approach to its campaign. Up until the end of July 2013 over 320 attacks had taken place inside the Yemen, according to figures produced by *IHS-Jane's*. If things were to continue at this rate nearly 550 attacks would occur in 2013. This total would almost equal the total number of terrorist attacks in the country up to the end of 2012.

To make such a turnaround in a single year is quite a feat and risks placing the Yemen into a situation very like what is happening in Pakistan. For the first time in history the rate of terrorism attacks in the Yemen has exceeded that in Pakistan on the basis of number of attacks per head of population. This is one of many indicators that can be used to look at the prevailing security situation in a specific country. In Pakistan 1,084 attacks were reported by *IHS-Jane's* up to the middle of July 2013. Given its population of 176.7 million in contrast to the Yemen's 24.8 million, the 320 attacks over the same period in the Yemen is a very worrying trend.

One specific individual who was also targeted by the United States in the Yemen was Anwar al-Awlaki. He was a United States citizen who became a radical preacher and took up residence in the Yemen. He was reported to be heavily involved in the magazine known as *Inspire* that is published by AQAP. He was killed by an armed UMA strike on 30 September 2011. Two weeks later his 16-year-old son was also killed in a similar strike in the Yemen. The death of Awlaki's son and also Al-Shihri are unlikely to mark the end of armed UMA strikes in the Yemen.

Analysis of Relationship Between Armed UMA Attacks and Strikes Against NATO Tankers in Pakistan

For the Taliban avenging the armed UMA strikes against their colleagues in the North-East of Pakistan poses a problem. Who do they retaliate against? Attacks on the Pakistani security forces who the Taliban can allege are colluding with the Americans offers one approach. Attacking NATO forces just over the border in Afghanistan provides another. However, there is a much more visible way of exacting revenge for the deaths of their colleagues and any civilians who have been caught up in the violence. That is to attack convoys of NATO tankers as they move fuel supplies from the maritime hub in Karachi overland into Afghanistan.

Figures released by NATO show that around 100 tankers cross the border from Pakistan into Afghanistan every day. They carry an average of 13,000–15,000 gallons of fuel. Afghanistan has no indigenous reserves of fuel, so every litre required by NATO has to be brought in from overseas. They are accompanied by another 200 trucks carrying other supplies.

Due to the terrain in the area the convoys have little freedom of manoeuvre when it comes to choosing alternative routes. Two major crossing-points have been in regular use. One is Torkham on the Khyber Pass and the other is Chaman in Balochistan. With such restricted routes, the tankers provide easy prey for Taliban fighters on both sides of the border. Tactics vary from intimidation of drivers through executions of those captured at the wheel of a NATO tanker to laying IEDs on the roads with the aim of killing those involved and also destroying the cargo. In one attack to the north of Kabul in July 2012 at least twenty NATO fuel tankers were blown up at an overnight staging post.

While such events do occur quite frequently with a total of 283 attacks having been reported by the South Asia Terrorism Portal over the period from 2008 to the end of July 2013, a valid question to ask is whether there is any correlation with armed UMA strikes in Pakistan.

Year	NATO Tanker Attacks	UMA Strikes
2008	8	41
2009	25	54
2010	100	129
2011	113	79
2012	20	53
2013	17	16
Total	**283**	**372**

Table C.1: The rate of attacks on NATO tankers in Pakistan

Looking at Table C.1 from a superficial viewpoint there does appear to be some degree of correlation between the two sets of figures. They are not different by an order of magnitude and aside from the slight departure in 2011 when the UMA strikes go down and the NATO attacks slightly increase, the overall trends are broadly similar. Indeed, it is possible to suggest from the figures that the attacks on NATO tankers do follow a broad trend allied to the rate of armed UMA strikes.

Analysis of a detailed month-by-month comparison actually reveals periods of time where it would be possible to suggest that the attacks on NATO tankers follow a profile that would be expected if a tit-for-tat strategy was being adopted by the Taliban. The correlation between the two profiles is 0.64; a number that implies some form of causal relationship between the two data sets.

However, a deeper look into the figures reveals some other factors that need to be considered. The armed UMA strikes in Pakistan mainly occur in areas away from the NATO tanker routes. The profiles of damaging attacks discussed in Appendix A show that even on the ground, retaliation does not simply occur at a local level. In some cases that would be very difficult as local security forces are often not present.

The detailed figures do show a time lag between the start of Taliban attacks on the NATO convoys and the increasing rate of armed UMA strikes.

Once they caught up, however, there did not seem to be any motivation to go beyond a strategy that is reminiscent of 'an eye for an eye'. This is an approach often adopted by Al Qaeda and one for which they have some theological justification in the way in which they interpret the Holy Quran. As armed UMA strikes peaked in Pakistan in 2010, the rate of attacks on NATO tankers levelled off. As armed UMA strikes decreased in 2011 there was a small time delay as the attacks against NATO tankers continued.

In November 2011 that all changed when Pakistan stopped NATO using its territory to re-supply its forces in Afghanistan. This was in retaliation for an attack mounted by United States forces on two military outposts in Pakistan that cost the lives of twenty-eight Pakistani soldiers. It was a new low point in the already troubled relationship between the Pakistani government and the United States.

In December 2011 there were no armed UMA strikes in Pakistan as the United States government sought not to fan the flames of an already difficult situation. The hiatus lasted for a few months. The disruption of the routes created a huge problem for NATO as for the previous ten years they had provided the means by which 75 per cent of the ammunition and foodstuffs used in the campaign had been brought into Afghanistan. During that time NATO had to open supply routes from the north into Afghanistan. These were much longer and yet paradoxically safer until the tankers crossed into Afghanistan.

For Pakistan, however, the suspension of the supply routes was always going to be a temporary measure as the country could not afford to lose the income generated from the movement of the fuel. Reports emerging at the time suggested that Pakistan was asking for $1M a day as a fee. In return NATO requested greater security as the trucks transited the country. Once Pakistan restored the supply routes in May 2012, the insurgent attacks on NATO tankers and re-supply convoys soon resumed.

Of course, the picture that emerges from the data could be misleading. The implication is that the Taliban hierarchy actually takes steps to attack NATO tankers. Given their resources, maintaining such a campaign is not difficult. It is not resource-intensive. Groups based near to the supply routes would not find it hard to keep up attacks. That is fine for groups closely allied to the Taliban based in the North-West Frontier Province areas. However, NATO tankers also operate on routes that avoid those areas, preferring to journey through Balochistan on a more direct route into Afghanistan from Karachi.

Year	Balochistan	FATA	NWFP	Misc.	Total
2008	2	2	4	0	8
2009	14	3	8	0	25
2010	64	16	12	8	100
2011	60	37	10	6	113
2012	9	10	0	1	20
2013	8	7	2	0	17
Total	**157**	**75**	**36**	**15**	**283**

Table C.2: Geographic distribution of attacks on NATO tankers [Source: South Asia Terrorism Portal]

Table C.2 shows the geographic distribution of attacks on NATO tankers. It shows that over half the attacks took place in Balochistan. This is a place where no armed UMA strikes have ever been conducted by the United States. It is a region with its own specific insurgency problem. Yet it is the area where attacks on NATO far outstrip anything in the FATA or NWFP.

The figures in Balochistan mirror the shape of those for the UMA strikes. Peaking at sixty-four in 2011, the figures tail away as the armed UMA strikes decrease from their peak illustrated in Table C.1. By October 2011 over 254 vehicles had been burnt in what at the time was estimated (fairly accurately) to be 112 attacks. Eleven of those took place in Quetta, the provincial capital of Balochistan.

All the attacks took place in ten of the twenty-six districts of Balochistan with the majority occurring in the Qalat division which is in the centre of the province. Here ninety NATO tankers were torched in sixty separate attacks since NATO intervened in Afghanistan. Across the whole of Balochistan fifty people had also been killed and thirty-six injured up to the end of September 2011. The figures were confirmed by political sources when in 2011 the Home Secretary of Balochistan noted in a statement that 136 NATO tankers had been destroyed in fifty-six attacks resulting in thirty-four deaths and twenty-three people being wounded. Being a NATO tanker driver was clearly a hazardous occupation.

What does Table C.2 suggest? Despite the obvious correlation between the rate of attacks on NATO tankers in Balochistan it would be hard to suggest that somehow the Balochi insurgency was in some way exacting revenge on behalf of their brothers who were under attack in the sanctuaries of North-West Pakistan. The objectives of the two insurgencies are somewhat different. The Balochis have long sought to create an independent Balochistan. Their motivations for attacking NATO convoys are unlikely to be directly related to the objectives of the Taliban in the north-west of the country. What is more likely is that the Balochis see the attacks on NATO tankers as a relatively simple and low-cost way of highlighting the deficiencies in the Pakistani security situation in Balochistan.

What these figures do not reveal is the degree to which the various routes have been prioritized by NATO. Given the results, it seems likely that NATO have been biased towards sending the majority of their fuel through the shortest possible route through Balochistan. The other routes up to the Khyber Pass are longer and hence provide more opportunities for the convoys to be intercepted. They are also through the more remote areas of Pakistan where the central government has less control. Deals to potentially buy local security from militias in the transit areas are unlikely to provide the kind of security needed. Some may even have been reneged upon by local people who in the past had been happy to be paid to provide security for the convoys as they transited their areas.

Bibliography

Axe, D., *Warbots* (Nimble Books, 2008)

Benjamin, M., *Drone Warfare: Killing by Remote Control* (Versobooks, 2013)

Bowman, M., *Mosquito Missions: RAF and Commonwealth De Havilland Mosquitos* (Pen & Sword, 2012)

Cooksley, P.G., *Flying Bomb* (Robert Hale, 1979)

Cull, B., *Diver! Diver! Diver!: RAF and American Fighter Pilots Battle the V-1 Assault over South-East England 1944–1945* (Grub Street, 2008)

Gordon, Y., *Soviet/Russian Unmanned Aerial Vehicles* (Midland Publishing, 2005)

Hinsley, F.H. and Simkins, C.A.G., *British Intelligence in the Second World War: Volume 4* (Cambridge University Press, 1990)

Johnston, P.B. and Sarbahi, A.K., *The Impact of US Drone Strikes on Terrorism in Pakistan and Afghanistan* (RAND Corporation, 2013)

Jones, R.V., *Most Secret War: British Scientific Intelligence 1939–1945* (Penguin Books, 1978)

Klaidman, D., *Kill or Capture: The War on Terror and the Soul of the Obama Presidency* (Houghton, Mifflin, Harcourt, 2012)

Lashmar, P., *Secret Spy Flights of the Cold War* (Sutton Publishing, 1996)

Longmate, N., *The Doodlebugs: The Dramatic Story of the Flying Bombs of World War II* (Arrow Books, 1981)

Martin, M.J., *Predator: The Remote Control Air War over Iraq and Afghanistan: A Pilot's Story* (Zenith Press, 2010)

Newcome, L., *Unmanned Aviation: A Brief History of Unmanned Aerial Vehicles* (Pen & Sword, 2004)

Nijboer, D., *Meteor vs V-1 Flying Bomb, 1944* (Osprey Publishing, 2012)

Ogley, R., *Doodlebugs and Rockets: The Battle of the Flying Bombs* (Froglets Publications, 1992)

Singer, P.W., *Wired for War: The Robotics Revolution and Conflict in the 21st Century* (The Penguin Press, 2009)

Sloggett, D.R., 'Anatomy of an Insurgency: Are Unmanned Aircraft Strikes Fuelling Terrorism?' (ISMOR Workshop at Royal Holloway University, London, July 2013)

Sloggett, D.R., 'Briefing to the Royal United Services Institute on UMA' (London, September 2013)

Williams, A., *Operation Crossbow: The Untold Story of Photographic Intelligence and the Search for Hitler's V-Weapons* (Preface, 2013)

Yenne, B., *Attack of the Drones: A History of Unmanned Aerial Conflict* (Zenith Press, 2004)

Yenne, B., *Birds of Prey: Predators, Reapers and America's Newest UAVs in Combat* (Speciality Press, 2010)

Zaloga, S., *V-1 Flying Bomb 1942–1952: Hitler's Infamous 'Doodlebug'* (Osprey Publishing, 2005)

Index

INDEX